Global Collaboration:

Neuroscience as Paradigmatic

ROBERT HENMAN

FOREWORD BY PHILIP MCSHANE

AP

Axial Publishing
Vancouver

Printed in Canada by
Grandview Printing Co. Ltd.
Vancouver, Canada
Email: info@grandviewprinting.com

Axial Publishing
www.axialpublishing.com

Canadian Cataloguing in Publication Data
Henman, Robert 1949 –
Global Collaboration
ISBN   978-0-9780945-9-1
1. Neuroscience   2. Functional Collaboration   3. Philosophy

Text layout and cover:
James Duffy and Christina Ghorayeb

To my grandchildren, Jake, June and Josie

"Would it not be nice to see, and be seized by seeing, the
sunflower in the seed."[*]

---

[*] McShane, Philip, (2013) Futurology Express, Axial Publishing, Vancouver, BC, page 14.

Contents

Global collaboration is upon us, whether we are greedy or gracious. The greedy in these past centuries were better at global reach than the gracious. There is the obviousness of this from the twentieth century emergence of powerful multinational corporations side by side with feeble nongovernment organizations: it is easier to exploit land and labor on a grand scale than to foster local well-being.[1] Henman's problem is to arrive at "a resolute and effective intervention in this historical process,"[2] in which, so far, it would seem that "people who are doing the most harm are doing it and the people who could do the most good are not."[3] I think of what is called strip mining on steroids wiping out mountains being battled by pockets of legal intervention, and thus I present you with an awkward implicit criticism: are NGO's or little pockets of cultural and legal resistance among those that I and Henman consider "people who could do the most good" who are failing? The implicit criticism is a nice nudge towards thinking of the relative helplessness of care and graciousness in the face of the present dominant bent of the controllers of the historical process. However, I leave the nudge there, leaving it to Henman to weave round it as part of his thesis on progress in his Epilogue.

The quotations at notes 2 and 3 are from Bernard Lonergan, on whose achievements Henman's work leans. Based on his digestion of that achievement, Henman moves to tackle the operations of the neurosciences as they are presently pursued and culturally absorbed. He does so in an elementary and suggestive fashion in the body of the book. He does so by gallantly venturing out of his usual fields of inquiry: child studies and education. Why he does so I leave to him to discuss in his Introduction.

---

[1] One can, of course, think of local successes, like Erin Brockovich-Ellis stand in 1993 against the Pacific Gas and Electric Company of California. Further, it is worth thinking of the portrayal of that achievement in the 2000 film *Erin Brockovich*, in terms of the effect of aesthetic presentations of saner human ways. Bernard Lonergan talks of the importance of art (*Insight*, 209; *Topics in Education*, 232) for cultural change, but realistically it is, unsupported, a feeble nudge. Tchaikovsky's *1812 Overture* does little to stop lunatic military marches across continents. We need Lonergan's larger Overture (see below pp. 42-3), the task of *Method in Theology* 250, lines 18-33, a task I have quaintly named his *1833 Overture*.
[2] Bernard Lonergan, *Phenomenology and Logic*, CWL 18, 306
[3] *Ibid.*, p. 307.

However, what is neat about his effort is that, in his search for a way into a new global control of meaning, he has hit on a zone of turmoil that is significant, the zone of neurodynamic research and technology.[4] I throw in here some oddments of the context of that significance that compactly express the reach of Henman's work. So, we may think of the neurodynamics of the business personality, where business as mentioned there can be viewed in a broad manner that includes the business of law, the businesses of local, national and global government, the businesses of war, the business of the Catholic Vatican or other religious groupings.

For Henman the significant zone that he turned towards was the business of psychology,[5] and his focus in the body of the work is on the underpinning of the field that comes from the chemistry of mind but flows, in uppers and downers and diets and multitudinous other ways, into our stumbling cultures. For that reason, I find it useful to weave in here some broader musings of Bernard Lonergan.

If we are to twist subtly into a saner globe,

> the first difficulty is psychological. The static phase is a somber world for men brought up on the strong drink of expansion. They have to be cured of their appetite for making more and more money that they may have more and more money to invest and so make more money and have more money to invest. They have to be fitted out with a mentality that will aim at and be content with a going concern and standard of living. It is not an easy task

---

[4] One of Henman's achievements is to point to the concrete meshing of neurodynamic studies with the movements of historical change. A central interest of his is the possibility of the serious Christian contribution of Bernard Lonergan to such shifting. One might usefully muse over past blockages to progress that are associated with unbalanced neurodynamics. The quotation to follow from Thom Hartmann, *Walking the Blues Away*, Par St Press, 2005, 40, serves as a symbol of the symbiosis of recurrence-schemes of cranial realities with history's patterns.
"Within a generation of the widespread introduction of literacy in Europe, the Catholic Church and its Inquisition murdered over a million women as 'witches' … With left-hemisphere dominance spreading across the culture, the men rose up and took over in a brutal and bloody way."

[5] The disorientations of present neuroscientific method are spelled out in Henman's second article. The naïve realism of the researcher leaves her or him trapped in a broad nominalism with no suspicion that one needs to be rescued.

to effect this change, for, as the Wise Man saith, the number of fools is infinite.[6]

A static phase of the economy is mentioned early in that quotation. It is a phase in the dynamics of economic creativities where a massive complex cluster of innovative technology levels out, awaiting the next creative surge.[7] But here one must tune into a full heuristic of long-term historical progress in provision to see more clearly and soberingly the weave of millennia to come: in the long run the ratio of the flow of primary business to the flow of capital provision business is to be an increasing quantity.[8] However, what this means is beyond present statement in our undifferentiated world of money-flows and business ventures. The neurodynamics of that undifferentiated mess holds thinkers regarding it quite captive.[9]

There are, then, deeper psychological issues regarding the dominance of finance and its massively muddled thinking.[10]

---

[6] *For a New Political Economy*, edited by Philip McShane, University of Toronto Press, 1998, Collected Works of Bernard Lonergan, volume 21, 98. Cited below as *For a New Political Economy*.

[7] The problem of the missed division of financial flows is thoroughly dealt with in the work cited in note 2. How might I intimate it briefly here? You may note, in a pause about the shops and stores in your own block, that there are two types of firm, those that provide consumer goods and those that provide the means of providing those goods. Without the development of this distinction, from late school texts on, we will remain in the dastardly present mess, not only of a well-paid economic pseudo-science, not only of the neurotic parasitism of financiers and stock-traders, but in the increasingly unbreathable air of global ecological disasters.

[8] Contemporary discussions of global inequalities—an obvious reality grimly visible in the public domains of first, second and less developed spots of the globe—lack the simple scientific basis to detect the source of such insanitary insanity. See, e.g. the popular work by Thomas Piketty, *Capitalism in the Twenty-First Century*, translated by Arthur Goldhammer, Harvard University Press, 2014. On the flaws in that thinking see Philip McShane, *Piketty's Plight and the Global Future: Economics for Dummies*, Axial Publishing, Vancouver, 2014.

[9] This issue points one into that complex neurodynamic area, in its infancy, of e.g. the chemo dynamics of the superego: here the community of scientists and historians of science need to follow Henman's suggested path of cyclic functional collaboration.

[10] I skim along here, citing Lonergan's work. Henman has more to say on the topic in his Epilogue.

Progress cannot wear blinkers; so, if we have stressed the excellence of the exchange economy, we must also be at pains to determine its defects. A fundamental defect lies in the innocent first step of the solution, in which those who are willing to contribute for little or no return are brushed aside, to make the exchange system an exclusive club for businessmen. With the psychological effects of this arbitrary procedure, we are quite familiar. It produces the split personality of the businessperson in his office and the respected citizen in his home. It turns out the pure types of the uplift worker who cannot get down to business, and of the common cynic who takes a business view of larger issues. But these are symptoms of a deeper malady.[11]

The deeper malady concerns the neurodynamic units of humanity, persons that are less mature and less adapted to the dynamics of history: within that spectrum of the less fit for survival there is the helplessness hunger of little persons and the vacant eyes of the neurodynamically or organically dysfunctional. Do related images bubble up in your neurochemicals to nudge your reading to tones of sympathy or damp eyes? However, that bubbling is far from effectively global: it can "tumble down the Niagara of fine sentiments and noble dreams."[12] Where, though, might this bubbling go?

So we arrive at Henman's problem and the subtlety of his modes.

Philip McShane

---

[11] *For a New Political Economy*, 35–36.
[12] *Ibid.*, p. 36

INTRODUCTION

Professor McShane has compactly pointed to the full reach of my work, and to that, I shall turn later in this Introduction, but in fullest fashion in my Epilogue. It seems best here to start modestly with the zone into which I venture in the body of the book, granted that you already have a suspicion from the Foreword of a larger set of overlapping contexts. There are three chapters there that present three articles in which I attempted to communicate three different aspects of Lonergan's thought as they apply to neuroscience.[1] The topic of the third and final article is the more relevant one regarding what is required to bring neuroscience into line with the *new science*.[2] That article is elementary in terms of explaining the dynamics of functional collaboration. It is a description of the order of functional collaboration with the added feature of a particular experiment in neuroscience. The article also does not touch on the many difficulties involved with getting functional collaboration started or implementing generalized empirical method.[3] This is the task of our times. Preliminary efforts at beginning need be the focus at this stage of history. Just what those difficulties are and how to overcome them is the topic of the Epilogue.

---

[1] I attempted to offer in these three articles what Lonergan outlines in his Fourth lecture of The Larkin-Stuart Lectures of November 15, 1973; *The Scope of Renewal*. Lonergan, B, (2004) *Philosophical and Theological papers 1965–1980*, CWL 17, University of Toronto Press, p. 293. Section 3.1 "Assimilation of the New." "...attending to their performance, (meaning the scientist) figuring out what is involved in any process from inquiry through discovery to experimentation and verification, and assembling the elements of the larger movement from one discovery to another." Parentheses added.

[2] I place these terms in italics to emphasize that the *new science* is not operational. It is embodied in an image of the order of functional collaboration as a way to order the content of the sciences.

[3] There were only two responses to my articles. Interestingly, one from Anuj Rastogi at the Institute of Medical Science for Collaborative Program in Neuroscience at the University of Toronto, expresses the need for a more intelligible order. "What is needed at this point is a broader theoretical framework in which to contextualize the plethora of scattered data." http://www.crossingdialogues.com/Ms-D14-03.pdf (2014) p. 68. Rastogi is correct and functional collaboration would intelligently order such data. Dr. Terrance Quinn offered a response raising the question, "Is it needed?" in a positive context. http://www.crossingdialogues.com/Ms-D14-01.pdf.

But let me talk of the difficulties that related to my venture into the topics of the three chapters. My problem was to keep the focus relatively restricted in these, even though the broader view lurked over all three essays, and was more difficult to avoid in the third article. I had two audiences in mind in my elementary venture: the usual audience of the journal, those interested in the goings-on of neuroscience, and the audience that is someway tuned to Lonergan's work. The latter audience was a consideration for this book-format: my initial effort was an outreach to those working in the neurosciences, although I knew that the Lonergan audience was a fringe presence.

For the audience in the field of neuroscience, and indeed, particularly for the small audience that is in control of the content of the journal in which these essays appeared, there was need for a cautious diplomatic approach. The moves I was suggesting for neuroscience were and are quite strange to the community of searchers in the area. I kept in mind, in my effort, the work of Professor McShane, *Randomness, Statistics, and Emergence*, that was originally a doctorate thesis in Oxford University.[4] There was the need to write directly, but with subtlety, to an audience quite unfamiliar with self-attention. So, one has to weave in rather lightweight hints at the identification of the key variables that are only cloudily recognized in the conventional research zone. The cloudy recognition is scarcely more than a refinement of commonsense meanings of words like sensing, attention, thinking, deciding, etc. The problem was, and is, to get the readers to link the reach for larger meaning both to themselves and to the data-zones of their normal scientific attention. In shifting from journal essays to this book I, so to speak, took a leaf out of McShane's work and avoided revisions that would block the readers of the initial audience. After all, these readers did, and do, represent the vast majority of the present population of readers.

This brings me to comment on the second audience, those who are students of Lonergan. Central to my work in the essays, and indeed to my effort here, is a nudging of that community. There is the obvious nudging towards devoting serious time to a personal climb into some science, and in the present case, it is a nudging towards neuroscience. It is, I would claim, simply not enough to devote themselves to self-attention on the basis of commonsense experience or even the experience of doing contemporary philosophy and theology in the manners of its present conventions. I would

---

[4] McShane, Philip, (1970) *Randomness, Statistics and Emergence*, Gill, MacMillan, and Notre Dame, 1970. Chapter Two of this booklet is largely my interpretation of McShane's Chapter Nine of that work.

claim that that was the message of Bernard Lonergan when he wrote the following.

> By classicism I mean the fruit of an unsuccessful education in which, first, there is no real grasp of theory of any kind – mathematical, scientific, philosophic, or theological. Theory is proposed and studied, but in the subject there is no real serious differentiation of consciousness; all we get as a theory are the broader simplifications offered by a professor to introduce or round off a lecture or a course, or the products of *haute vulgarisation*. But he is never bitten by theory; he has no apprehension, no understanding, for example, of the fact that Newton spent weeks in his room in which he barely bothered looking at his food, while he was working out the theory of universal gravitation.[5]

The first two chapters here express the need for the central focus of chapter three. How will neuroscience, or any science or scientist, rescue itself, her or himself from the vicissitudes of a modernity fraught with a world of science unable to meet the challenges of, not just our present age, but all the ages to come? McShane, in his Foreword, raises that reality in his discussion of the disarray of present economics. Over the past three decades, I have taught courses to students on educational methodology, ethics, more recently peace and conflict studies and child studies. The disarray McShane and I speak of dominates all these zones, and more, of inquiry and teaching. "A civilization in decline digs its own grave with a relentless consistency. It cannot be argued out of its self-destructive ways, for argument has a theoretical major premise, theoretical premises are asked to conform to matters of fact, and the facts in the situation produced by decline more and more are the absurdities that proceed from inattention, oversight, unreasonableness and irresponsibility."[6] The historical process is fundamentally awry and the transition out of this Axial Period is in need of a process that reorders our scientific efforts in line with the emerging order of the structure of human cognition.

---

[5] Lonergan, Bernard, (1996) *Philosophical and Theological Papers, Collected Works of Bernard Lonergan*, volume 6, University of Toronto Press, 155.

[6] Lonergan, Bernard, (1973) *Method in Theology*, Darton, Longman & Todd, London, 1972, 55.

Though leaves are many, the root is one;
Through all the lying days of my youth
I swayed my leaves and flowers in the sun;
Now I may wither into the truth.[7]

[7] Yeats, W.B. (1968) *W. B. Yeats: Selected Poetry*, Ed. by A. Norman Jeffares, "The Coming of Wisdom with Time", page 45.

CHAPTER 1

CAN BRAIN SCANNING AND IMAGING TECHNIQUES CONTRIBUTE TO A THEORY OF THINKING?

Introduction

This article is a brief analysis of both the current methodology practiced in neurocognitive research and a problem inherent in that methodology. The analysis investigates the problem by: 1) adverting to the distinction between self-observation and reflection on the performance of test subjects, 2) identifying a problematic assumption within the practice of neuroscientific research and 3) providing pointers on possible lines of progress highlighting problems possibly entailed by the mentioned assumption.

There are different techniques presently used in neuroscientific research (Iliescu and Dannemiller, 2001, p.133). The following are six conventional ways of gathering data in the neurosciences: 1) electroencephalography technique (EEG), 2) MRI scans, 3) fMRI scans, 4) near-infrared spectroscopy (NIRS), 5) positron emission tomography (PET), and 6) magneto-encephalography (MEG).

Neuroscientists are attempting to "map" the brain and to determine the functions of different areas of the brain and the interactions among them. By doing so, neurocognitive scientists are attempting to develop a theory of thinking. In turn, neuropsychologists are studying relationships between the functioning of the brain and human behavior (Henman, 2000), as well as searching for the breakdowns responsible for diseases such as schizophrenia, autism and Down's syndrome. The six techniques listed above gather data that reveal activity between brain locales that correspond to conscious operations and cognitive experiences. Correspondence is established empirically by measuring the simultaneous or sequential occurrences of mental acts and the brain activities. Verification of these correlated events is achieved by the repetition of the experiments. There are differing types of data generated by the different techniques of *mind mapping,* but the data are similar because they are technically produced images or scans of cerebral activity. The data from the first four techniques consists of graphs and images that signal the occurrence of electrochemical changes during mental and conscious activity. PET and MEG research are designed to record changes in chemistry and magnetic fields occurring in the brain while test subjects perform designated tasks.

Tests are developed to evoke specific mental operations (paying attention, puzzling, memory, reasoning, decision-making, planning, speaking) and conscious states such as emotions and moods. One purpose in pursuing more specific and detailed descriptions of cerebral activity is to determine the causes of certain brain disorders linked to genetic mutations. It is hoped that such studies may lead to the prevention of some disorders as well as a better understanding of the genetic development of the human brain (Nelson, 2001 p. 149).

Current neuroscience experimentation

A common approach in neurocognitive research is to focus first on a particular mental operation or human experience of which the researcher wishes to find the cerebral correlate. As an example, let us say that we are searching for the cerebral correlates of conscious acts of problem solving. The presupposition is that, if a researcher can locate the region or regions of the brain that manifest synaptic activity while a test subject is problem solving, this information will aid in understanding both problem solving and the links between what neurosciences uncover and what cognitive psychologists are studying. To these ends, researchers design experiments for test subjects in order to obtain the data of graphs and scans signifying the cerebral correlates of problem solving. In clinical research, cognitive acts in participants are frequently stimulated through problem solving. For example, Figure 1 below is a puzzle, which could be utilized to locate the cerebral correlates of the different cognitive operations occurring when a person is problem solving.

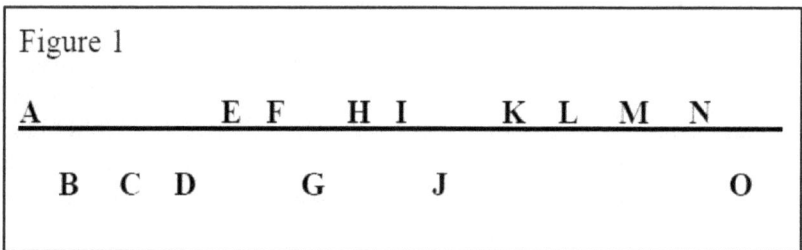

Figure 1									
A		E F	H I		K	L	M	N	
B	C D		G	J					O

In fact, I have been using this puzzle (Figure 1) in my classes on cognitional theory and ethical decision-making for over 25 years. The problem to solve is why some letters are on top while others are on the bottom. I add a second question to the exercise. The students are asked to reflect on their conscious operations in trying to solve this puzzle. This is a leisurely exercise in which the students are not to help one another and are to relax with the process. The diagram provides the first set of data or clues. They attend to the diagram, the image. Now, a non-standard part of the problem is to notice that attending is also a datum. Of course, I do not mean that the experience of 'attending' is data that can be measured, but rather is data in the sense of being experience, being something that can be described and explained. Students are wondering and puzzling, revealing a second experience of data, in response to my question: What law or formula is in play that arranges the letters in the above format? Therefore, we can describe patterns of mental acts that occur: paying attention and puzzling.

Some of the students "get it" quite quickly; they experience an insight, a third mental operation that is a further datum. Others are somewhat slower in achieving insight, and some do not get it at all. I provide the following hint (Figure 2) for those who cannot get it. This also serves as verification for those who, at this point, think they have it and often assists others who have not achieved the insight.

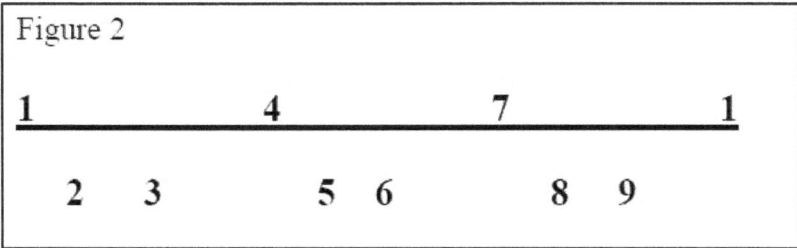

Figure 2

Some students do not experience the insight without assistance, and eventually I ask one of the students who thinks she has it to explain why the letters are separated in the above manner. When I try to draw the students' attention to the second question (i.e., how are they operating in trying to solve this problem), it is very difficult for them to shift their attention away from the wanted solution and to focus on their mental operations. If they cannot attend adequately to their own acts of thinking, and this is often the case, I assist with further clues labelling and suggesting what their cognitional acts might be. The data needs to be generated first by their desire to solve the puzzle and then by their actual performance. Once all the students have understood the law that is functioning in the puzzle, I then focus directly on the second question.

How did you solve the puzzle? What was going on in you as you were solving the puzzle? Eventually a percentage, usually about half of the class, begins to notice and to acknowledge a distinction between the content and the cognitive acts. Let me now focus on the term "problem-solving". From where do cognitive neuroscience researchers derive this terminology? What empirical data are researchers referring to when they use the term *problem solving*? Are they referring to any empirical data when using the term *problem solving*? We use terms such as paying attention, thinking, puzzling, explaining, understanding, knowing, judging, problem-solving, decision-making and planning in commonsense conversations, in philosophy and in science, but there may be no explicit empirical references cited. Neurocognitive scientists theorize that by locating the neural correlates of these "terms" they will achieve a better understanding of the meaning of these terms. Since the terms originate as first-person reports, scientists hope to shift the meaning of the terms into a third-person perspective, so that they can be studied less "subjectively". Hence, they are interested in locating observable data as potential correlates of these first-person reports. However, are the images and graphs that visually record synaptic activity occurring in the organism similar in any manner to the human subject's conscious experiences of attending, puzzling, explaining, understanding, judging, knowing, planning or decision-making? The data of mental acts and the images from brain scans are two distinct forms of data. They are related but distinct. Even though the listed terms signify, in a non-explanatory manner, mental operations, they are judged to be subjective reports, and we commonly use the terms as if we have some data-reference *in mind*. Do we? How can such terms be used in a scientific context or experiment if we have no specific empirical data-reference?

I suggest the following reflection on performance as a means to *resolve this problem*. What I am referring to here is seldom adverted to and yet is implicit in normal procedure. In fact, I am inviting attention by the researchers, on the processes of the researchers themselves. The researcher has a problem she *desires* to *resolve*. The researcher *desires* to *know* what the cerebral correlates for *problem-solving* are in the brain? The researcher's first task is to *design* an experiment that he or she *believes* will achieve the *intended* outcome. So, in *question* form: *What form of experiment is required?* The researcher *puzzles* and *reflects* on various possibilities, *reviews* other experiments by other researchers, until eventually an *insight* occurs (an "ah ha" experience). This or that particular experiment should provide the outcome that the researcher is *seeking*. However, the researcher does not *know* at this stage. She may *search* more literature on similar studies or set up mock experiments in an attempt to *verify*

4

her former *insight*. *Certainty* need not be the goal, but she may *feel sufficiently convinced* that the form of experiment she has *settled* on is *reasonable*. The only way to *verify* that this confidence is warranted is to run the experiment.

Let us retrace the steps of the researcher's problem-solving (Steps 1 to 9 from Benton et al., 2005, p. 67–71; Steps 10 to 21 from Lonergan, 2001, p. 322–323):

1. Desired outcome: Locate organic correlates of problem solving. (Data)

2. A *What question*: What form of experiment will achieve this outcome?

3. *Insight* (Ideas, understanding of possible designs)

4. *Formulation* into a *Concept*, a formulated *Answer* to a *What* question

5. *Will it work?* Seeking verification of *insight*

6. *Indirect insight*

7. *Judgment*. Reasoning

8. *Planning*. How to set up and run the experiment?

9. Further *what* and *Is* questions

10. Set up and run the experiment. Doing

11. *Yes,* it worked. Verification

12. Outcomes achieved

13. *What-to-do* with the outcomes?

14. Further *insights*

15. *Develop* options

16. *Choose* the *most reasonable* option. *Judgment*

17. *Decision* to implement option

18. *How* to implement the option

19. Further *insights* on how to implement option

20. *Planning* the implementation

21. *Implementing* the option.

The listing in sequence of the first seven acts of problem solving corresponds to the cognitive acts that my students report experiencing, when solving the alphabet puzzle. The terms in italics refer to mental acts that researchers can notice in themselves when trying to solve a problem. In other words, these distinct mental acts function heuristically as a dynamic sequence or pattern of acts occurring in problem solving. As such, they are the basic elements of a theory of thinking. Verification of these elements requires that researchers reflect on their own performance while solving a problem and ask this *question* while doing so: Do we experience these acts and in this order when we are solving a problem? There are three possible judgments: "Yes", "No" or "Maybe". Such judgments are mental acts and can be made explicit by attending to one's own performance during problem solving. Even a judgment of No requires the same process as a Yes, and, even though both are *reasonable* responses to any **IS** question, in this particular case, **No** is nonsensical. The researcher has to go through the first seven steps in order to verify the order and number of her own mental acts and arrive at the judgment: "No, I do not perform these acts when I am problem-solving". Why is this negative response nonsensical? This answer requires reflection on cognitive performance. Doing so will reveal the same patterned acts of attending to whatever is puzzling, asking questions, surmising what answers might work, asking whether they in fact work, checking them out and arriving at judgments.

Concerns about the reliability of self-observation are not groundless. One reflects on one's performance, and such reflection is quite different depending on whether one is reflecting on one's emotions or on one's scientific procedures. Brentano concluded that experiencing emotions and thinking yield the same type of data (Brentano, 1874/2013). Even so, there is no conflict, and no lack of distinction. Anger is not a mental operation. It is an emotion or conscious state, but the data of distinct types of cognitive acts are better classified as operations, not states. It is generally held that emotions are affective rather than apprehensive. The mental operations are generated by the desire to understand, but emotions are often evoked by either the inability to understand or a refusal to understand (Lonergan, 1992, p. 219). Anger, it would seem, has its origins in the complex integral dynamic of the human chemical, psychic and intellectual makeup of a person in response to an experience (Lonergan, 1992, Chapter XV, section 7). Brentano, one would assume, reflected on his own performance in order to arrive at his conclusion. Was he observing himself or reflecting on his own performance? If not, what procedure did he employ to obtain and verify his conclusions?

What does this brief description of cognitive acts offer the neuroscientist? In preliminary fashion, it sketches the elements for a theory of cognition that later uses of imaging and scanning techniques can differentiate and track. It assists the neuroscientist in refining his or her own tests so that test-subjects can be deliberately "walked through" the various mental acts and thus reveal with more specificity the brain locales and activities that are correlates of the distinct types of acts.

What I am suggesting here is something that will add to present progress. The addition will lead to a theory of cognition that includes attending to one's own performance, to the data of one's own mental acts, when engaged in problem solving. First-person reports of these acts can be followed by research into how each type of act has correlates in neural activities at specific brain locales. The goal is greater understanding of these mental acts and their neurochemical antecedents, all the while recognizing that working out such correlations is not the same as identifying the acts with their correlates.

I have offered only a listing and minimal descriptions of mental acts. An explanatory account of their functions and relations to one another would require a much larger work. Efforts have been made to understand acts of insight and to work out a procedure for achieving insight into insight and insight into other cognitive acts (McShane, 1975, Ch. 3 and 4; Lonergan, 1992, Ch. 3 & 4). Steinberg and Davidson's *The Nature of Insight* (Steinberg and Davidson, 1995) presents the work of 25 psychologists developing a descriptive phenomenology of the act of insight and its relationship to problem-solving and thinking. The text is limited in its expression by a lack of distinction between description and explanation and by the supposition that concepts precede insights. However, in the Preface, Janet Metcalfe makes a statement at odds with this assumption:

Qualitatively, then, this kind of model has the right feel for the emergence of concepts as a result of insight (Steinberg and Davidson, 1995, p. xiii)

This statement can be verified in one's own experience. Concepts are the result of insight, not the reverse (Lonergan, 1970, pp. 38, 42). Though the above quotation is not followed up with any explanation, it challenges the supposition inherent in conceptual analysis, which is dominant in present scientific procedure. The author was probably not aware of the paradigmatic shift she was advocating by making such a statement. Moreover, that shift is one of meaning. Steinberg's and Davidson's text does not do justice to the acts of intellect in the manner that Lonergan offers in his text *Insight* (Lonergan, 1992), but it does point beyond the reductionist tendencies (a residue of nineteenth-century positivism) often present in neurocognitive literature.

If a theory of cognition cannot be achieved solely through imaging and scanning techniques, what outcomes can these techniques produce? While a theory of cognition needs to account for the data produced by scanning and imaging research, must it not also include and account for the data produced by the researcher and that of the subject's performance? Both accounts are important if the research is to be thorough and the resulting theory comprehensive. Still, understanding the relationship of the brain to body chemistry, psychology, observable behavior and mind remains an ongoing challenge.

A detailed account of the mental acts increases the specification of locales and events in the brain. Understanding more completely the relations between the various types of mental acts will help to influence the design of future experiments. A specific puzzle may be the reason why different combinations of brain locales are sometimes activated during the occurrence of the same type of mental act. In contrast to these varying combinations, scans may record different rates of synaptic activity in the same cerebral regions for the same type of mental act during different problem-solving experiments. This is not to suggest that the mental acts "cause" the synaptic activities or that the latter "cause" the former. A cause is an explanation, and a comprehensive explanation will treat correlates as dependent variables without relegating either to an epiphenomenal status (Lonergan, 1992, p. 316–318).

Neuroscientists have to work out the specific outcomes of cognitive neuroscience research. However, a more complete account would result from "framing" research projects including all the relevant data (neurological and psychological). This contribution will arrive, eventually, at a comprehensive theory of thinking (Gulyás, 2009, p. 142).

There are doubts within the neuroscientific community, as to whether or not a theory of cognition can be achieved through the current techniques. In the Introduction to *Neural Correlates of Thinking*, the editors, Kraft, Gulyás and Poppel highlight this problematic, and Gulyás raises the same question in his article, "Functional Neuroimaging and the Logic of Conscious and Unconscious Mental Processes." He asks; "Are these techniques helping us reveal the neurobiological underpinnings of cognitive processes?" (Gulyás, 2009, p. 142). He does not ask what these "cognitive processes" are, but at least he accepts a distinction between the two. What follows are some comments by a few leading neuroscientists, regarding their doubts about their methodology (from Bandettini, 2009, p 31–32):

What about thinking? A major theme in this book is the quest to understand thinking. The question that most reading this chapter will want to know the answer to is: "What can fMRI, or more generally, neuro-imaging, contribute to our pursuit of an understanding of thinking?" Does it really help to be able to look into the brain? To borrow an analogy, can one really truly understand how computers work by opening up a computer chassis and probing the components with a heat gun? Can identifying the when, where, and how much in the brain provide enough information so that we can begin, from this information, to derive principles of thinking? Even if we had a perfect picture at infinite spatial and temporal resolution of what was actually happening in the brain during thought, would we even then begin to understand thinking? Does it really matter what the limits of fMRI are with regard to answering questions about thinking?

It seems apparent that to truly understand the brain, a much wider context (physical and evolutionary factors) needs to be considered. Thinking itself might someday be deconstructed into simple algorithms that can be carried out within different media other than brains. Perhaps a simple model of interacting layers of neuronal networks may emerge as being able to explain thought (Hawkins and Blakeslee 2004). It is my feeling that because thinking is a subjective process, it tends to be shrouded in mystery, and potentially elevated to a status, either correctly or incorrectly, that defies understanding.

At the end of the day, we might be able to then say that x network, on x spatial scale, is directly related to say, theory of thinking, willed action, and humor. So fMRI reveals the functions of specific processing modules. Does this really tell us anything that will help our understanding of thinking? Do we need to know what modules overlap in function or how large they are or where they are located in the brain?

Does this information really matter? What spatial scale in the brain is the most critical for the understanding of thinking? While all of our tools are able to probe many different spatial scales, there are also many which have not been investigated yet. Does this matter?

Horace Barlow and Rita Carter add emphases to this quandary.

...reductionism is limited because its drive is to look for explanations at lower levels in the organizational tree. ...Can we learn about the mind in the same way that we might seek to understand a machine—by taking it apart and examining its parts?

Neurophysiologist Horace Barlow believes this approach can bring about important insights but can never tell the full story (Carter, 1998/2000, p. 43)

With the help of [imaging techniques] can we exploit the differences between conscious and unconscious brain processes? (Gulyás, 2009, p. 142)

It still remains unclear whether it is justified to assume that neural assemblies are actually the basic units of cognition (Öllinger, 2009, p. 75)

...a coherent theory of thinking is lacking...a book exclusively dedicated to...gaining insight into the process...seems warranted (Kraft et al., 2009, p. 6)

The connection between neuroanatomy, neurochemistry, and neurodevelopment, and the behavioral research in cognition are rather tenuous (Nelson et al, 2001, p. 415)

As always, an understanding of the mind must guide the search for its neural underpinnings (Nelson et al, 2001, p. 429)

Richard Moodey (personal communication, email correspondence at: lonerganl@googlegroups.com, 2013) offers an interesting insight into the relationship between researchers and human subjects in the following:

> When working with human subjects, the neuroscientist has to ask people about their experiences in order to get information that he cannot know immediately, and relate this to his observations as an "outsider."

> His outsider observations are aided by ever more sophisticated apparatus, but the connections with the phenomenological accounts of the research subjects are what give fuller meaning to the external observations.

Moodey's point is that both first-person reports of test subjects and the third-person reports of researchers are legitimate sources of data. Will this more inclusive perspective make researchers' accounts more comprehensive? The problematic that obfuscates the settling of the issue stated in the previous quotations is expressed summarily by Lonergan:

> In this fashion, intelligence is reduced to a pattern of sensations; sensation is reduced to a neural pattern; neural patterns are reduced to chemical processes; and chemical processes to subatomic movements. The force of this reductionism, however,

is proportionate to the tendency to conceive the real as a subdivision of the 'already out there now'. When that tendency is rejected, reductionism vanishes (Lonergan, 1992, p. 282–283)

Part of the problem then, is how to unify results. The verifiable patterns of mental acts are conscious events, conscious in the sense of experienced. At the same time, neuroscience is uncovering verifiable patterns of aggregates of biochemical and cellular events. However, the conscious acts of attending, puzzling, understanding and judging "look" nothing like scanned images or graphs (McShane, 2013).

Generalized Empirical Method

To reach a more complete and balanced account of mental acts, researchers can reflect on their own performance when they, for example, are problem solving. For, performance, such as problem solving, provides data about mental acts. To identify functional relations can be part of formulating a structured heuristic, a framework for studying thinking and knowing. The following quotation describes one procedure for developing a more balanced framework.

> Generalized empirical method [hereafter GEM] operates on a combination of both the data of sense and the data of consciousness: it does not treat of objects without taking into account the corresponding operations of the subject: it does not treat of the subject's operations without taking into account the corresponding object (Lonergan, 1985, p. 141).

The procedure goes forward by first listing distinct types of operations, then by describing them, and eventually by grasping the functional relations among them. Without this procedure, the scanning and imaging techniques presently operative in the various neurosciences will probably overlook some of the first-person data and so not achieve a comprehensive theory of cognition. GEM can provide more complete and balanced accounts of mental acts to be correlated with cerebral locales and events during experimentation. For example, the detailed listing of cognitive acts can serve as a guide for tracking components in problem solving. It will also provide researchers with a standard model of performance of her or his research procedure. It is a standard model of performance in research not unlike the standard model of the periodic table. The intelligibility of the periodic table has provided a foundation for progressive development in pharmacology, industry and medicine for the past 144 years. The standard model of the mental operations of the human mind is paradigmatic in that it is the source of the very nature of a paradigm that results from reflection on performance. Margaret Masterman supports this in her comment on Kuhn:

> Kuhn's form of thinking…reflects the complexity of the material. … because he has taken a close look at what mathematicians really do…" (Masterman, 1970, p. 60)

Kuhn, even though he advocates reflection on the performance of the mathematician, does not expose the operations of the mathematician's mental operations (McShane, 1980, p. 5f.).

A standard model of mental operations results from reflecting on the very operations that one is engaged in while trying to develop such a model. The practice of GEM and a self-appropriation of one's own mental acts are steps toward systematic control over research, evaluations and recommendations as well as over any ontogenetic account of the development of the cerebral organism (McShane, 1980, p. 49 f.).

Only by performance and reflection on that performance can one come to notice the distinctions and differentiations involved in problem-solving. Generalized empirical method calls into question the traditional definition of human subjectivity. How does a researcher know if he or she is working with the correct data, asking the right question(s), getting the correct insight(s), and formulating that insight(s) into a correct judgment(s)? The researcher does so by gathering all the relevant data, asking all the relevant questions and answering them (McShane, 1975, chapter 4). In other words, objectivity is achieved through authentic subjectivity (Lonergan, 1992, chapter 13; McShane, 1976, pp. 120–126). There is an objective element to each mental act. In other words, how does the researcher know when she is asking the right question? In short, because intelligence seeks to be intelligent and that drive is the human desire to understand correctly. In the final judgment of yes or no, one achieves a greater degree of objectivity. When one verifies an insight, one is not verifying data, one is verifying an **insight into data** and such an act looks nothing like the "seen" data (McShane, 2013). Such distinct operations ground GEM and scientific research in general. The larger issue is whether science is an effort to understand data, all data. As long as science refuses to attempt to understand the data of human consciousness and first-person reports relevant to them, scientific research will remain truncated in its development.

> The neglected subject does not know himself. The truncated subject not only does not know himself but also is unaware of his

ignorance and so, in one way or another, concludes that what he does not know does not exist (Lonergan, 1974, p. 73).

Conclusion

There are further questions and doubts raised in the literature about the process and method of the present scanning and mapping techniques of the brain. While these concerns and doubts are raised by some of the leading researchers, it seems that their theories are ignored in much of the literature. In particular, there is the present restriction to experimentation through scanning and imaging techniques. A more complete method, though, will lift these excellent results into a balanced account that includes detailed description of conscious events. In other words, a Generalized Empirical Method (GEM). Such doubts and concerns are indicative of an observation, by the authors referenced above, in order to distinguish between two forms of data. Scanning and imaging techniques contribute to our understanding of the human brain and its relationship to human genetics, psychology, mental acts and intellectual development. An explanatory account of the mental acts of the researcher occurring when performing experimental research can broaden the perspective of the researcher. Might this could be of help to cognitive neuroscience in order to achieve what it has been seeking for some time, namely, an understanding of the brain in its relationship to the data of mental operations?

My response to the question expressed in the title of this essay is obviously "No." A theory of mental operations cannot be provided through scanning and imaging techniques but such techniques can and do assist in understanding the underlying biological aspects of cerebral activity. I envisage, then, a shift from reductionist tendencies to a more balanced methodology. In addition, this shift would also provide the beginnings of a new heuristics for phylogenetic and ontogenetic development of the cerebral organism and its relationship to a theory of thinking (Lonergan, 1992, p. 489, McShane, 1975, p. 106).

References

Bandettini PA. (2009) "Functional MRI Limitations and Aspirations". In: Kraft B, Gulyás B, Pöppel E. (Eds) *Neural Correlates of Thinking*. Springer, Heidelberg: 15–38.

Benton J, Gillis-Drage A, McShane P. (2005) *Introducing Critical Thinking*. Axial Press, Halifax.

Brentano FC. (1874/2013) "On the Distinction between Self-observation and Inner Perception". Dial Phil Ment Neuro Sci, 6:4–7.

Carter R. (1998/2000) *Mapping the Mind.* University of California Press, Berkeley, Los Angeles and London.

Gulyás B. (2009) "Functional Neuroimaging and the Logic of Conscious and Unconscious Mental Processes". In: Kraft B, Gulyás B, Pöppel E. (Eds) *Neural Correlates of Thinking.* Springer, Heidelberg: 141–174.

Henman R. (2000) "Judgment, Reality and Dissociative Consciousness". Method: Journal of Lonergan Studies, 18, #2:179–186.

Iliescu BF, Dannemiller J. (2001) "Brain-behavior Relationship in Early Visual Development". In: Nelson CA, Luciana M. (Eds.) *Handbook of Developmental Cognitive Neuroscience.* MIT Press, Cambridge, MA: 127–146.

Kraft B, Gulyás B, Pöppel E. (2009) "Neural Correlates of Thinking". In: Kraft B, Gulyás B, Pöppel E. (Eds) *Neural Correlates of Thinking.* Springer, Heidelberg: 3–14.

Lonergan B. (1992) *Insight: A Study of Human Understanding.* University of Toronto Press, Toronto.

Lonergan B. (1974) *A Second Collection,* Edited by W. Ryan & B. Tyrrell The Westminster Press, Philadelphia.

Lonergan B. (1985) *A Third Collection.* Paulist Press, New York.

Lonergan B. (2001) *Phenomenology and Logic: The Boston College Lectures on Mathematical Logic and Existentialism.* University of Toronto Press, Toronto.

Lonergan, B. (1970) *Verbum: Word and Idea in Aquinas.* University of Notre Dame Press, Notre Dame, IN.

Masterman M. (1970) "The Nature of a Paradigm". In: Lakatos I, Musgrave A. (Eds) *Criticism and the Growth of Knowledge.* Cambridge University Press, Cambridge: 59–90.

McShane P. (1975) *Wealth of Self and Wealth of Nations: Self-axis of the Great Ascent.* Exposition Press, New York. NY.

McShane P. (1976) *The Shaping of Foundations.* University Press of America, Washington, DC.

McShane P. (1980) *Lonergan's Challenge to the University and the Economy.* University Press of America, Washington, DC.

McShane, P. (2013) Posthumous 3: A Commentary on the Inside http://www.philipmcshane.org/

Öllinger M. (2009) "EEG and Thinking". In: Kraft B, Gulyás B, Pöppel E. (Eds) *Neural Correlates of Thinking.* Springer, Heidelberg: 65–82.

Steinberg R, Davidson J. (1995) *The Nature of Insight.* MIT Press, Cambridge.

CHAPTER 2

GENERALIZED EMPIRICAL METHOD: A CONTEXT FOR A DISCUSSION OF
LANGUAGE USAGE IN NEUROSCIENCE

Introduction

In a paper on Generalized Empirical Method (Henman, 2013), a distinction was drawn between the data of sense and the data of consciousness. That distinction laid the groundwork for an affirmation of a theory of knowing as a conscious activity that can be empirically acknowledged through reflection on one's scientific performance (Henman, 2013, p. 51, Lonergan, 1970, p. xiii). This article appeals to that distinction as a context for a discussion of particular language usage in neuroscientific literature that attributes mental acts to biological processes. In doing so, an unintentional neglect of the data of consciousness is perpetuated, as well as a denial of the empirical nature of conscious acts or states. This also contributes to undermining the possibility of a more adequate understanding of biochemical processes. Discussions of a) objectivity, b) knowing as a conscious activity and c) the biological process of evolutionary theory will provide further contexts towards a shift in methodology providing the possibility of a better understanding of the relationship between the cerebral organ and consciousness.

To provide a context for the above discussions, what follows is a quotation of Bernard Lonergan's definition of Generalized Empirical Method and his position on reality and knowing.

> Generalized Empirical Method operates on a combination of both the data of sense and the data of consciousness: it does not treat of objects without taking into account the corresponding operations of the subject: it does not treat of the subject's operations without taking into account the corresponding object (Lonergan, 1985, p. 141).

Lonergan's position on knowing is referred to as critical realism. He outlines in his book, *Insight: A Study of Human Understanding* (Lonergan, 1992) the intellectual acts of knowing as experience, question, insight, and judgment. [1] These acts can be acknowledged, at least experientially, by

---

[1] Interpreters of Lonergan's cognitional theory have equated his thought with that of idealism. Such an interpretation misses the point concerning the role of judgment. Judgment is the result of a verified insight into data and it is only in such

adverting to our own acts of conscious operations, or to our reflection on our performance when involved in either common sense knowing or theoretical knowing. This process he calls self-appropriation (Insight, 1992, p. 13–17). Lonergan also establishes that this process of knowing is how we know reality. The real is the verified. Verification is the establishment of an unconditioned or a verified insight. Reality is a correctly understood experience (Lonergan, 1992, p.123, 230, 278. Lonergan, 2004, 126–127. McShane, 1975, p. 42). In a correct judgment, we know reality. Experiencing and the subsequent operations of the intellect are components of both knowing and reality. As a whole they constitute a heuristic structure leading to knowledge and reality. Lonergan's cognitional theory, epistemology and generalized empirical method provide a context for our discussion on objectivity and its relationship to reductionism. These discussions provide a methodical foundation for understanding the relationship between the brain and consciousness. We begin with a quotation from Lonergan which describes the problem of reductionism in the following manner.

> In this fashion, intelligence is reduced to a pattern of sensations; sensation is reduced to a neural pattern; neural patterns are reduced to chemical processes; and chemical processes to subatomic movements. The force of this reductionism, however, is proportionate to the tendency to conceive the real as a subdivision of the 'already out there now'. When that tendency is rejected, reductionism vanishes (Lonergan, 1992, p. 282–283).

Language Usage Data

The discussion begins with a focus on the use of language that mediates a reductionist tendency in meaning in neuroscientific research. That usage is not a pure form of reductionism in as much as reductionism assumes that what is being reduced is known at least empirically. In the case of conscious acts, they are presently unintentionally neglected and therefore empirically unacknowledged by the scientific community as data to be understood.

What follows are specific terms that refer to mental operations but, in the context exhibited below, are attributed to biological processes. They are

---

a judgment that knowledge is achieved. Insight is concrete and is the pivotal act towards knowing but without raising the IS question; Is it so?, nothing is known as verified. Such errors of interpretation result from an insufficient reflection on performance when one is involved in theoretic knowing.

as follows: **information, decoding, storage, memory, determine, represent, recognize, process,** and **knowledge**. The following quotations contain examples of such usage taken from a current text in cellular and molecular neuroscience (Byrne et al, 2014).

"...with respect to the type of **information** transferred to the neuron" (Byrne, 2014, p. 4).

"...is responsible for transmitting **information**. This **information** may be primary...or processed **information**..." (Byrne, 2014, p. 5).

"...are correlated with the type of **information** processed by the particular neuron..." (Byrne, 2014, p. 5).

"...how **information** is processed in the cerebral cortex" (Byrne, 2014, p. 8).

"...DNA **information** is copied into messenger RNA..." (Byrne, 2014, p. 149).

"How does DNA **store** genetic **information**?" (Byrne, 2014, p. 150)

"...the **information stored** in genome can be transmitted to new cells..." (Byrne, 2014, p. 150).

"...transcription and translation are induced and required for **memory storage**" (Byrne, 2014, p. 162).

"...biophysical and biochemical properties as well as the ways in which neurons are connected to each other to **process information and generate behavior**" (Byrne, 2014, p. 591).

"However, in most other examples of memory, it is considerably less clear how the **information is retrieved**. This is especially true in memory systems that involve the **storage of information** for patterns, facts, and events" (Byrne, 2014, p. 623).

"...V4 is thought to **represent** various aspects of the form and identity of visual objects" (Knierim, 2014, p. 567).

"...a downstream neuron can **determine** whether the odor was apple or cherry + apple by decoding whether PN2 fired on all 3 cycles of the LFP or on only the second and third cycles" (Knierim, 2014, p. 570).

"...thus a neural system that has **prior knowledge** about the probabilities of such inputs...will successfully **decode** the image..." (Knierim, 2014, p. 575).

"...the decoder system may **recognize** that it is highly unlikely that an elephant is in the living room ..." (Knierim, 2014, p. 575).

"... based on its anatomical connectivity with the eyes, we can deduce that the lateral geniculate nucleus (LGN) of the thalamus **represents** visual **information**" (Knierim, Byrne, 2014, p. 563).

I have bold-faced particular terms in the above quotations to focus the reader's attention on the possible meaning of these terms. I have obtained samples from one main text but perusal of other texts in neuroscientific literature reveals similar usage. Can the operations that the terms **interpret, determine, knowledge, recognize, decode, information, recognize** and **formulate** refer to be empirically located in cellular processes? Do cellular processes have these abilities? The following reflection will focus on the term **information**.

What constitutes information?

I have cited 14 uses of the term **information** in the sampling provided from the referenced text. There are many more. This is sufficient for our purposes. The Oxford English Dictionary (O.E.D.) defines information as facts or knowledge provided or learned. A definition serves its purpose only if the recipient of the definition understands the various terms and the relationships between those terms. If there is no understanding the phrase is data to be understood. So, what is a **fact**, what is **knowledge** and what is it to **learn**? The bold-faced terms need to be understood if this definition is to be helpful in our discussion. If an adult says to a child of age 2 that a snow storm is coming tomorrow and the child does not know what the terms, snow, storm and tomorrow mean, is it **information** for the child? What is missing is an understanding of these three terms in the child's mind. For the child the words are data to be understood. The statement becomes **information** only once the child understands the terms and how the meaning of each term relates to the meaning of the other terms. Such understanding is learning and knowledge and the occurrence of the storm has the potential to become a fact for the child. In the same manner the definition of **information** becomes **information** once the individual terms are understood.

What constitutes information are conscious operations resulting in correctly understanding data. Until such understanding occurs, what is named **information** in the samplings is in fact data. The sampling offered above uses the term **information** as an attribute of cells and neurons that is passed on from one cell or neuron to another cell or neuron. Do the cells pass on correctly understood data? Think of a mechanism such as an internal combustion engine being made up of many parts, each part functioning with the other parts to achieve one overall function, moving an automobile from one place to another. The overall purpose is achieved because each part works integrally with the other parts. But each part does not hold the overall design in its particular design or function. The design is in the mind of the designer and the designer has an image of the overall pattern and seeks out the parts that will function together to achieve that overall function. Shifting this analogy to the biochemical processes of cells and neurons we have processes each with a function, but no overview of the design of the human anatomy or the brain is innate to each cell or neuron. Biochemical processes do not pass on information any more than the parts of an engine pass on information to the other parts. With this understanding in mind, is the use and attribution of the meaning of the term **information** to biochemical processes an inhibition to understanding the data of biochemical processes that are occurring (McShane, 1970, p. 187)? This question requires further methodological points regarding objectivity and emergent probability which follow.

What is Objectivity?

It will be helpful to provide a quotation from Charles Osgoode, an experimental psychologist.

> What is the environment? The answer to this question comes promptly—the environment is "out there". It is this book, the walls of this room, the people passing to and fro; it is everything that is outside of us. This answer is of course a rational one, but it is founded upon an elaborate system of inferences developed through a lifetime of experiencing. If a forefinger is placed along the lower ridge of the eye socket so that its tip is against the nose and the other eye is covered, pressing the eyeball gently and moving it up and down will cause the environment to jump back and forth. Now this is manifestly unreasonable! Any force sufficient to shake the room would also have been felt as vibration. But what, then, is the explanation of this phenomenon? (Osgoode, 1953, p. 1)

The key term in this quotation is explanation. It is, in fact, an explanation that is required. A correct explanation is achieved when an insight has been verified. An explanation is a cause, or a cause is an explanation. Once an explanation has been verified one has 'created' knowledge or an understanding of experience. Reality and knowledge are instances of correctly understood experience. There is no objective reality "out there" beyond the human mind. And if there were, one could not somehow get outside oneself in order to have some super look. And even if one could, and there was something "out there", he or she would still be faced with the question; What is it? raising again the issue of objectivity.[2] There are further experiments on experiential objectivity that one can carry out, such as the image changing when removing glasses. Which image is "out there"; the hazy image or the clear image? Can the hazy image be "out there"? If not, where is it? (McShane, 1975, chapter 5) The Nobel recipient in 1967 for physiology and medicine, H. K. Hartline, over a period of twenty years showed in his experiments with the Limulus, the horseshoe crab, that visual images originate in the photoreceptor cells (Hartline, 1967). John Dowling states that more recent work has confirmed Hartline's position (Dowling, 1992, p. 210).

A second quotation from Knierim; "...the lateral geniculate nucleus of the thalamus represents visual information (Knierim, p 563)." Where is this visual image that the LGN of the thalamus represents? Do these processes represent some "other" reality or are they, in fact, processes of emerging complexity manifested as a visual experience in consciousness? Are such processes representatives of anything?

A quotation from Bernard Lonergan on objectivity situates the role of data in human knowing.

> If objectivity is a matter of elementary extroversion, then the objective interpreter has to have more to look at than spatially ordered marks on paper; not only the marks but also the meanings have to be 'out there'; and the difference between an objective interpreter and one that is merely subjective is that the objective interpreter observes simply the meanings that are obviously 'out there,' while the merely subjective interpreter 'reads' his (or her) own ideas 'into' statements that obviously possess quite a different meaning. But the plain fact is that there is nothing 'out there' except spatially ordered marks; to appeal to dictionaries and

---

[2] Understanding this existential shift through scientific explanation resolves issues surrounding Husserl's numenology and Kant's noumenon.

to grammars, to linguistic and stylistic studies, is to appeal to more marks. The proximate source of the whole experiential component in the meaning of both objective and subjective interpreters lies in their own experience; the proximate source of the whole intellectual component lies in their own insights; the proximate source of the whole reflective component lies in their own critical reflection. If the criterion of objectivity is the 'obviously out there,' then there is no objective interpretation whatever; there is only gaping at ordered marks, and the only order is spatial (Lonergan, 1992, p. 605).

The "out there" experience of spatiality is a result of the extroverted character or quality of animal/human consciousness (Hartline, 1967). Now, if sense data are not "out there", then the visual images, sounds, smells, tastes and touches are in fact integral patterns of our brain chemistry. [3] Brain scans and graphs are more of the same.

The point to this discussion is that such usage as provided in the beginning of this section is misleading and contributes to bolstering up the reductionist tendency. We grow as children in the commonsense horizon developing a "sense" of reality as what is solid and out there. As Osgoode and Lonergan remind us, there is more to the picture. Objectivity is not achieved by looking at the data; objectivity is achieved by verifying one's insight into data and that may be as simple as just verifying that there is data. Such verification is a judgment and understanding is still in play. We cannot escape the need for understanding, but we can be led into and remain in the commonsense mode of reality throughout our lives, even if we are a scientist. McShane offers the following statement on the need for an adequate personal position on objectivity.

"It is clear that one cannot handle the senses with scientific clarity without an accurate personal position on objectivity and extroversion." (McShane, 1976, p. 93)

The commonsense horizon (Lonergan, 1973, p. 235) is what we all live in and move about in every moment. So, for the person of common sense, reality is the "already out there now real", but for the scientist, reality is a

---

[3] Lysergic acid diethylamide (LSD) experiments manifest that visual images and other sense data are often distorted. Just where are these images, sounds, tastes, or smells? How can the human brain distort sense data if it is "out there"?

verified explanation and not some subdivision of the "already out there now" (Lonergan, 1992, p. 523). Reality, or the real, is known in the judgment reached through the heuristic combination of the conscious cognitive acts. These acts constitute knowing and reality, and both are reached by correctly understanding experience or data. Objectivity is the product of authentic subjectivity, asking and answering all the relevant questions until an insight or explanation is verified.

When these two different horizons are not adequately distinguished there is a mixing or meshing of common sense and theory in scientific research (Lonergan, 2004, 10-11. McShane, 1975, Chapter One: Horizons. Clayton, Zanardi, 2014, p. 19). So, the neuroscientist works at gathering data, describing structures and processes of cells, determining functions and relationships of functions between cells and more. Without distinguishing between the two notions of reality, the order and expression of the above procedures become meshed in such a manner that the researcher is moving in and out of the two different horizons unaware that he or she is doing so (Lonergan, 1992, p. 318–324). Shoppa and Zanardi have explored this in their latest work and offer the following on this meshing of horizons in relationship to development. "That history is ongoing, development is incomplete, and the proof lies in the mixture of descriptive and explanatory terms in almost every science." (Shoppa, Zanardi, 2014, p. 40) They add in a footnote the following to the above statement; "In the neurosciences this mixture appears with some frequency in common references to 'mechanisms in the brain' and 'communication of information' along the neural pathways." (Shoppa, Zanardi, 2014, p. 40)

In view of the lack of distinction between the two horizons, the reductionist inclination is reinforced and becomes an inhibition to establishing a theory of thinking adequate to explain the relationships between the cerebral correlates and the acts and states of consciousness, and finally the distinction between the two horizons. Such distinctions are achievable through empirical reflection on one's own performance when doing science (Henman, 2013).

Abnormal development occurring in the cerebral organ can inhibit the development of intellectual potential. The activities of the cerebral organ are an underlying level to conscious activity (Lonergan, 1992, p. 543). The point is that development in understanding does not occur in biological processes, it occurs in the mind. The tendency to attribute such activities to occurring in biological processes is "a stumbling block to the appreciation of what precisely is going on in cell biochemistry" (McShane, 1970, p. 187). Development in understanding initiates changes in the underlying biological processes, e.g., synaptic activity and dendritic growth accommodating that conscious development in understanding. The human person functions as an integral subject whereupon higher and more complex levels of activity integrate less complex levels of activity. When there are distortions on lower levels, higher levels are affected in their functioning. All that we are is brought integrally forward into consciousness including any distortions or abnormalities.

The Function of the Cerebral Organ

James Knierim's offers the following statement on the function of the brain.

> The essential functions of the brain and nervous system are to collect **information** about the external world and about the internal state of the body; to **interpret that information**; to **determine how that information conforms** with the needs and goals of the organism; and to **formulate** an appropriate behavioral response (if necessary) to accomplish those goals. (Knierim, 2014, p. 563)

How did Knierim arrive at his statement? Knierim as a professor and researcher at the **Mind/Brain Institute** at John Hopkins University has what would be considered the necessary background to adequately speculate on the function of the brain. Reflection on his own performance would reveal that Knierim **decided** what **data** was **relevant, interpreted** that data, **determined** how that **interpretation** is to be organized and **formulated** the above statement. What was going on in Knierim's mind while he was writing this statement? First, he would have had a question relating to the essential functions of the brain (Byrne, 2014, p. 563). That question might have been in the form: What is the function of the brain? That question is a conscious act seeking understanding or an insight, also a conscious act, into the function of the brain. That insight(s) he would formulate into an answer and eventually decide to write the above statement. So, we have data: Knierim's previous understanding of brain processes. We have a question on the function of the brain, various insights and formulation of those insights. In order of occurrence: Attention to data, curiosity expressed in a question, insights into the intelligibilities revealed by intellect in the data and finally judgment, or rather attending, questioning, understanding and judging. These terms refer to distinct mental operations that occur in the human mind, having biological correlates occurring in the brain. Through the exercise of his own mental operations, Knierim "determined" the function of the cerebral organ, the nervous system, and if he reflected on his own performance he would be able to acknowledge his own mental operations. There is intimated a dualism in Knierim's statement through his attributing conscious mental operations to biological processes while he was carrying out such procedures as conscious operations. To which group of operations does he advert as a source of intelligence and what process does he use to make that decision? How would Knierim access those supposed biological operations? What empirical evidence is there to support Knierim's conclusions that biological processes carry out any of the operations that Knierim attributes to biological processes?

Biological processes function as schemes of recurrence within the parameters of chemical and physical laws and the randomness associated with the possibility of non-systematic events. The empirical occurrence of the cognitional acts as conscious, associated with the above terms, can be acknowledged by reflecting on one's own performance when involved in scientific work (Henman, 2013, p. 51).Through the conscious process of reflecting on one's performance, one arrives at the conclusion that knowing is a conscious activity and there is no evidence-based data to verify that cellular processes carry out such acts. The neurophysiologist, Sir John Eccles, raised the point at the Pontifical Academy of Sciences gathering in 1964 when he stated; "the ultimate reality for me is my conscious experiences, including perceptions, memories, dreams" (Eccles, 1964, p. 371). Eccles has worked extensively on attempts to understand the interface between the brain and the mind. Eccles, with a focus on acknowledging both brain and mind as separate entities and yet interfunctioning (Eccles, 1988), holds to that distinction. The distinction is not maintained in present research, where the approach is towards raising the chances of understanding the interface. Generalized empirical method will be a help in future work towards working out the interface.

If biological processes do not possess the ability to determine, what do they possess? There are patterns of intelligibility in data that the cognitive operations discover and seek to understand (Lonergan, 1992, p. 101). Without those operations nothing is known. There is just "looking" at fMRI scans, graphs and electron microscopic images. Knowing is a conscious activity. The biological processes are unconscious events. Once an insight is verified, the researcher has reached a judgment. A judgment is not "out there" in the data. A judgment is "in" the mind of the researcher. This is a curious expression if one is reading with presuppositions of positivism or the notion that objectivity is a subdivision of the "already out there now". In order to understand these biological processes better, I draw on Bernard Lonergan's theory of evolution which he calls emergent probability (Lonergan, 1992, p. 144–151). The following is a brief description of emergent probability from Lonergan's own writings.

> For the actual functioning of earlier schemes in the series fulfills the conditions for the possibility of the functioning of later schemes. As such conditions are fulfilled, the probability of the combination of the component events in a scheme jumps from a product of a set of proper fractions to the sum of those proper fractions. But what is probable, sooner or later occurs. When it occurs, a probability of emergence is replaced by a probability of survival; and as long as the scheme survives, it is in its turn fulfilling conditions for the possibility of still later schemes in the series. Such is the general notion of emergent probability (Lonergan, 1992. p. 145).

The above definition of emergent probability is descriptive and filling in the explanations will be the work of the various researchers in the fields and subfields of neuroscience. But for our purposes here it will serve to provide a contextual image for the following discussion of cellular and biochemical processes helping towards discovering the function of the cerebral organ.

If the brain is not the source of judgments, just what is the function of the brain? To understand the function of the brain we must first understand what is occurring along the neural pathways. This can be achieved by first understanding the cellular and chemical processes and secondly by determining the functions of these activities and then finally the relationship between the various functions (Lonergan, 1992, p. 489). The neural pathways do not *carry information* from a sensory site to the cerebral organism. So some transformation of cellular activity at the sensory site is initiated and this transformation in turn initiates a transformation of the biochemical cellular processes within the neural pathways. That transformation is a more complex level of functioning than that occurring at the sensory site. The change in cellular activity at the sensory site is a change in the normal scheme of cellular activity, a new pattern of activity occurs. The transformation initiated in the neural pathways is a change in its normal routine of activity. The transformation is eventually integrated by the functioning of the cerebral organism. All along the line we have data that consists of changes in biochemical activity and changing schemes of recurrence with different functions. There is no encoding of information on the axons in this activity. It is a series of transformations from different levels of activity that have functions and each level of functioning serves to initiate new schemes of recurrence with higher levels of functioning. And these functions can be determined by the mind's conscious operations of relating one level of activity to the former and succeeding levels of activity. These transformations consist of an increasingly more complex activity of cellular and biochemical processes and activity moving towards an integral finality that is a conscious experience and only experience. Lonergan describes briefly the process in the following quotation.

> ...chemical elements and compounds are higher integrations of otherwise coincidental manifolds of subatomic events; organisms are higher integrations of otherwise coincidental manifolds of chemical processes; sensitive consciousness is a higher integration of otherwise coincidental manifolds of changes in neural tissues; and accumulating insights are higher integrations of otherwise coincidental manifolds of images and data (Lonergan, 1992, p. 477).

The cerebral organism functions as an integral operator of biological processes. The cerebral organism is a higher level of activity that organizes neural pathway activity in a manner effecting awareness of activity: the human subject is conscious of an experience. There is an upward dynamism in which activity is reorganized by more complex systems so that what is cellular activity is experienced as sense data.

Where and how is the empirical evidence of these upward transformations of schemes of recurrence to be located? The intelligibilities discovered by intellect in data, through present neuroscientific work, provide sufficient evidence for these transformations. Knierim offers the following.

> Many neural circuits are fairly simple and are relatively well understood. These circuits may be little more than a sensory neuron synapsing with a motor neuron,..." a stimulus causes a sensory neuron to fire, which causes a muscle to twitch." ... These concepts are essential for describing more complex neural circuits that are formed from networks of thousands or more neurons (Knierim, 2014, p. 565).

This image is methodologically descriptive, not explanatory of the transformations. The image does serve to note, at least descriptively, an increasing intelligibility, but not intelligence, as we move from simpler to more complex levels of activity and functioning. An explanatory account of the transition from genetic composition to the synthesizing of proteins, to the formation of cells, to the emergence of organs and organisms, to conscious awareness, is a major challenge for neuroscience and evolutionists that will require a proper method of approach and the collaboration of many in the related fields before adequate results are forthcoming. There is a forward dynamism in biological processes towards the eventual awareness of experience. That dynamism is not "known" by biochemical processes. Unintentional reductionism is one of the hurdles to overcome, in order to provide the proper method required to understand this increasing dynamic complexity.

What is necessary for such a study is an understanding of two aspects of what is called genetic method (Lonergan, 1992, p. 484–506); development and emergent probability (Lonergan, 1992, p. 144–151), or what has been traditionally called a theory of evolution. By emergent probability I mean an explanatory account of related and emerging schemes of recurrence, not limited to natural selection or Darwin's notion of the survival of the fittest (McShane 1970, p. 233), that through the occurrence of a non-systematic event, establishes a new scheme of recurrence that functions as a new, higher and more complex level of activity (McShane, 1970, p 218–219). The new level of activity systematizes the non-systematic event emerging from the former level. The new level of activity is operative as a scheme of recurrence exhibiting a new function that is also open to new possibilities through the non-systematic possibility. The cell has its own particular scheme of recurrences, and cells group to form organs and organisms, and those organs and organisms have their own schemes of recurrence that have different functions then those of the cell. A new and higher level of functioning is occurring. In the same manner, the cerebral organ is a level of functioning through schemes of recurrence that admit and are driven to higher and more complex systems of functioning. This series of increasing complexity of activity and functioning find their final integral transformation as conscious experience. These transformations occur within the context of the physical and chemical laws functioning with statistical probabilities of the occurrence of a non-systematic event leading to higher levels of integral functioning. Research by Lonergan and McShane support this process as expressed in the three following quotations.

The non-systematic of one level is systematized on the next level without contradiction (McShane, 1970, p. 187).

For in the first place, an acknowledgment of the non-systematic leads to an affirmation of successive levels of scientific enquiry. If the non-systematic exists on the level of physics, then there are coincidental manifolds that can be systematized by a higher chemical level without violating any physical law. If the non-systematic exists on the level of chemistry, then on that level there are coincidental manifolds that can be systematized by a higher biological level without violating any chemical law. If the non-systematic exists on the level of biology, then on that level there are coincidental manifolds that can be systematized by a higher psychic level without violating any biological law. If the non-systematic exists on the level of the psyche, then on that level there are coincidental manifolds that can be systematized by a higher level of insight, and reflection, deliberation and choice, without violating any law of the psyche (Lonergan, 1992, p. 229–230).

"The function of the higher integration involves changes in the underlying manifold, and the changing manifold evokes a modified higher integration." (Lonergan, 1992, p. 495)

Over the past few decades, neuroscience has provided an enormous amount of data in mapping the brain, describing the many processes of cells and synaptic events, dendritic branch growth and their functions, leading to advancements in surgical and therapeutic procedures. An understanding of the general function of the brain and the acknowledgment of the mental acts as distinct data would increase the probability of development in understanding the biological and biochemical processes of the cerebral organ.

The fact remains that there is no empirical evidence to support Knierim's conclusions that biological processes carry out any of the mental operations that Knierim attributes to biological processes. Experiments such as Libet's work and others, noting cellular activity prior to one's conscious willing, are not empirically-based evidence of the brain "thinking" (Cibelli, 2012). These processes do not admit the experimenter to experience any particular acts. But there is empirical evidence to support the occurrence of mental operations in Knierim's mind.

The Function of Consciousness

What is the function of consciousness, and in this case, of human consciousness? The function of consciousness is awareness (Lonergan, 1992,

p. 344–346). Or more pointedly, consciousness is experiencing. But consciousness is more than a function. It is an awareness of the self. Explaining that relationship between the function of the brain and the function of consciousness will require that it be necessary to determine what consciousness is. A quotation from McShane provides methodical direction for that task.

> What is consciousness? The question is a massive empirical challenge of this millennium, moving up from the irritability of plants through higher levels of self-presence in plant and animal to the shades of human consciousness, where different consciousnesses of inquiry, judgment, planning and decision will be specified by investigating the chemical patterns of heterarchies of brain neurodynamics (McShane, 2011, p. 193, footnote 35).

The self-awareness of consciousness as our self-presence of existing is complemented by the expression of the heuristic sequence of cognitive operations as a solution to surviving in particular environments.

The Relationship between the Two Functions

We have two separate functions listed above of two distinct entities of the human person. The function of the brain is being an integrator of biological processes into possible intelligible patterns and being an operator towards higher schemes of recurrence (Lonergan, 1992, p 488–492). The function of consciousness is awareness of these patterns as well as the occurrence of cognition, the desire to understand discovered intelligibilities in data. The occurrence of correlates in the cerebral organ when conscious operations occur reveals that there is some form of relationship between these two activities or functions. What might that relationship be? The relationship between these two functions is that together they function as an integral collaborative solution to the survival of an organism within a particular environment "... data on the schemes of recurrence which include, say, reproduction on protozoa, lead the biologist first to the physics and chemistry of each scheme and further to the correlations which define a particular capacity for dealing with environment." (McShane, 1980, p. 68)

The task of future neuroscience is to explain that form of integral collaboration and to do that a transformation in methodology is required. Reductionism will only continue to inhibit this much needed transformation. The methodological pointers outlined above should provide the context for an understanding of the reductionist tendency of the other bold-faced terms listed in the third section of this essay. An adequate position on objectivity, generalized empirical method and emergent probability (evolutionary theory) outlined briefly in this essay would also contribute to answering Farina's question: "How is it possible to explain the infinite mystery and the paradoxes of human being(s) through brain mechanisms and physiological systems?" (Farina, 2014, p. 61)

Concluding Summary

In conclusion, I suggest a main inhibition to progress and development in neuroscientific research to be the lack of reflection on performance. To date, the immensity of the data and literature in the various fields and subfields of neuroscience is commendable, but it requires systematic organization. Generalized Empirical Method will provide new data to be considered that will eventually lift present methods of experimentation, observation and conclusion to a level that can meet the challenges of our times by shifting attention to a combination of the data of sense and the data of consciousness. The future will bring forth more refinements in technology that will surely provide even more exacting data on cerebral activity. This data will still require an adequate explanatory method in order to provide a genetic sequence of systematic development. That sequence of development can be initiated by the scientist's reflection on his or her own performance. There are clues in Lonergan's description of genetic method and development that will not only subsume the problems of procedure in scientific research, but also provide a method for gradually assisting in neuroscientific development (Lonergan, 1992, p. 489–504). Reflection on one's cognitive performance while experimenting needs to parallel the intellectual energy and resources that are presently dedicated to neuroscientific experimentation. The activity of such experimentation provides the operative conscious data for such reflection. The conclusions regarding the randomness of biophysical and biochemical events, the objectivity of statistical science and the fact of emergence reveal that reflection on method is essential to understanding the content of any and all science (McShane, 1970, p. 11).

References

Byrne, J., Heidelberger, R., Waxham, N. (2014) *From Molecules to Networks: An Introduction to Cellular and Molecular Neuroscience.* Third Edition, Academic Press, UK.

Cibelli, E. (2012) "Free will according to B. Lonergan and B. Libet", in Taddei-Ferretti, C. *Going Beyond Essentialism*. Napoli: Istituto Italiano per gli Studi Filosofici.

Dowling, J. (1992) *Neurons and Networks: An Introduction to Neuroscience*. The Belknap Press of Harvard University Press. Cambridge, MA.

Eccles, J. (1988) *The Brain-mind Problem in the Interactionist Approach*. Pontifical Academy of Sciences.

Eccles, J. (1964) *Proceedings of the 1964 Study Week of the Pontifical Academy of Sciences on Brain and Conscious Experience*.

Farina, G. (2014) "Some Reflections on the Phenomenological Method". http://www.crossingdialogues.com/Ms-A14-07.htm Dial Phil Ment Neuro Sci 2014; 7(2): 50–62.

Hartline, H. (1967) "Visual Receptors and Retinal Interaction". Nobel Lecture December 12, 1967. http://www.cns.nyu.edu/events/vjclub/classics/hartline-lecture.pdf

Henman, R. (2013) "Can Brain Scanning and Imaging Techniques Contribute to a Theory of Thinking?" DIAL PHIL MENT NEURO SCI 2013; 6(2): 49–56. http://www.crossingdialogues.com/Ms-A13-07.pdf

Knierim, J. (2014) "Information Processing in Neural Networks", in Byrne, J., Heidelberger, R., Waxham, N. (2014) *From Molecules to Networks: An Introduction to Cellular and Molecular Neuroscience*. Third Edition, Academic Press, UK.

Lonergan, B. (2004) *Philosophical and Theological Papers 1965–1980*, University of Toronto Press, Toronto, CA.

Lonergan, B. (1992) *Insight: A Study of Human Understanding*, CWL 3, University of Toronto Press.

Lonergan, B. (1985) *A Third Collection*. Paulist Press, New York.

Lonergan, B. (1973) *Method in Theology*, Darton, Longman & Todd. London, UK.

Lonergan, B. (1970) *Verbum: Word and Idea in Aquinas*, Edited by David Burrell, University of Notre Dame Press.

McShane, P. (2011) "The Hypothesis of a Non-accidental Human Participation in the Divine Active Spiration," in *Method: Journal of Lonergan Studies*. Vol. 2, No. 2

McShane, P. (1980) *Lonergan's Challenge to the University and the Economy*, University Press of America.

McShane, P. (1976) *The Shaping of Foundations*, University Press of America.

McShane, P. (1975) *Wealth of Self and Wealth of Nations: Self-axis of the Great Ascent*. Exposition Press, NY.

McShane, P. (1970) *Randomness, Statistics and Emergence*, Gill and MacMillan, Dublin, Ireland.

Osgoode, C. (1953) *Method and Theory in Experimental Psychology*, New York.

Shoppa, C., Zanardi, W. (2014) *Cracking the Case: Exercises in the New Comparative Interpretation*, Forty Acres Press, Texas.

CHAPTER 3

IMPLEMENTING GENERALIZED EMPIRICAL METHOD IN NEUROSCIENCE BY FUNCTIONALLY ORDERING TASKS

Introduction

The many tasks of neuroscientific studies are an evident feature of that broad field of inquiry and its application. In this essay, I shall enter into details of the divisions of such tasks in the present methodological suggestion of ways beyond present practice (Henman, 2013, 2015). However, it seems good to state, in the introduction, a fundamental root cause of present problems and disorientations (Henman, 2015, Clayton, Zanardi, 2014, p. 40). Disorientations relate to the intelligent ordering of tasks, which will become apparent in the later portions of this essay. I do so here in a very elementary descriptive fashion.

First, then, we can share the question, and indeed its common answers, "What is a science?" Common answers, we easily find, stem from the analysis of Aristotle of the nature of science. That analysis left the western tradition with a narrow and conventional view of good scientific work. My informed readers may think of other traditions, like those of China or North Africa, where serious sciences of health and building shook off that ill-fitting narrow perspective on the human effort to understand and control, even improve on, nature. There has been, in the West, a move towards what we might call clarity or ideal typology such that deviants from that clear type were considered, in some way, inferior. This is true whether one considers formulations of paradigms, such as constitute texts on philosophy of science, or discussions of paradigm-shifts, such as occur in the Kuhnian tradition.

Therefore, adverting, however briefly, to the position of Aristotle is extremely enlightening. Sir David Ross sums up my point: "Throughout the whole of his works we find him taking the view that all other sciences than the mathematical have the name of science only by courtesy, since they are occupied with matters in which contingency plays a part." (Ross, 1949, p. 14). This simple identification gives us a context for understanding what has happened, especially in the modern period, in reflections on science. Take, for example, the reflections that are the telling of the story of science, a story of contingencies spread over time and space. One does not think of such reflections as science. Much less does one think of applied science such as medicine, as science (Douglas, 2013).[1] What, then, of such topics as the teaching of science, the aesthetics of science, the ethics of science, the technologies of science? An Aristotelian perspective would have us consider such topics to be, as it were, outside the world of genuine science.

But are not these pursuits intelligent, rich with understandings of data, of the flow of events? "The immobility of the Aristotelian ideal conflicts with developing natural science, developing human science...," (Lonergan, 1972, p. 24) and with reflections on these developments, their application, their importance, their ethics and their aesthetics. There is, then, a whole world of *scientia* – in the normal sense of the translation from the Latin – that escapes from, or is denied dignity by, Aristotle's narrowness. It is time to raise the question, "What is science?" with a freshness that takes the full modern data seriously.

Discussions of Generalized Empirical Method (GEM) as a means of including both the data of sense and the data of consciousness in scientific research implicitly raise issues about just what that *new science* might consist in (McShane, 2013). This article will outline the structure of that new science which will be found eventually to function as a method of implementation of GEM (Henman, 2013, 2015).

The essay attempts to outline a method known as Functional Specialization. First, I quote Bernard Lonergan's definitions of a method: "Method is a framework for collaborative creativity" (Lonergan, 1973, p. xi). Later Lonergan expands on this definition to say that "method is a normative pattern of recurrent and related operations yielding cumulative and progressive results" (Lonergan, 1973, p 4) So, thinking of both definitions, would a normative pattern of related operations be a framework for collaborative creativity, and furthermore, would such activity provide cumulative and progressive results? To answer these questions I turn now to the data in the full reality of human interest in neuroscientific work.

Divisions of Labour in Neuroscience

This section of the essay will distinguish stages from data to results both in past, present and future experimentation in an effort to outline the "new science". In order to outline and describe the order and function of the divisions that occur during the scientific process from data to results, we

---

[1] Douglas' article goes to great length debating the dialectical history between pure and applied science. Reflection on scientific performance would end this debate, but such work is part of the process of a history of a developing science.

begin with a summation of a particular article describing an experiment in neuroscience.

The neurochemical basis of working memory (Furey *et al*, 1997)

Neurons projecting to the prefrontal cortex contain a specific neurotransmitter, acetylcholine, which is delivered by a specific type of neurotransmitter pathway, the cholinergic pathway. Recently, it has been hypothesized that an increase in acetylcholine to the frontal cortex might lead to an improvement in working memory (Furey *et al.*, 2000). Furey *et al.* administered the drug physostigmine (which increases the amount of acetylcholine available) to a small group of men and women who completed a working memory task as a functional Magnetic Resonance Imaging (fMRI) scanner monitored their brain activity. Participants completed the same task the following day but received a placebo (saline). The task involved watching a human face for three seconds, and then after a nine-second delay when the face was removed, identifying which of the two or more subsequently presented faces was that originally seen.

Participants who received the physostigmine showed increased activation in the visual cortex during the encoding of the face, activation that was significantly lower in the saline condition. The physostigmine condition also produced better face recognition when participants had to decide which of the faces had previously been presented. The finding suggests that the improvement in working memory may be due to enhanced visual processing in the earliest stages of encoding. One practical example of this may be to administer such drugs to patients with stark memory deficits, such as patients with Alzheimer's disease (Robbins *et al.*, 2000, p. 2275–76) (Martin, 2003, p. 167).

Functional Research

Before a researcher begins to plan a research project, one has a question within a hypothesis or a hypothesis. Once the question or the hypothesis has been formulated, the researcher works out what form of data will be required to test the question or the hypothesis, and the type of data required dictates the type of experiment required and the various technologies to be employed to provide the required data. Various meetings are held with colleagues to discuss the hypothesis and the form that an experiment might take.

An experiment such as the one described above requires a significant amount of planning before the experiment can occur. There are proposals, consent forms to ethics committees when using human subjects, applications for funding and/or grants need to be made, planning out the materials required, preparing images of faces, determining adequate times for recognition, selecting proper timing devices, staff required, facilities, recording technology, access to fMRI scanning technology, finding subjects, consent forms for subjects to sign, determining what information is to be conveyed to subjects prior to the actual conducting of the experiment, setting the date and place, plans for no-show subjects, approving and acquiring medications, interviewing subjects regarding their physical, psychological and mental health and suitability for the experiment, and finally, meetings with colleagues to monitor stages of planning to ensure all necessary planning is being carried out properly. This listing provides one with some idea of the immensity of conducting any research project.

Seasoned researchers would be familiar with this entire process. This process of planning can take months or a year or more before the experiment is finally conducted. Even though, in the present case, the experiment took place over two days, the post-experiment work can also take weeks, months or years to understand the data and what to do next.

Taking into consideration both the data of sense and the data of consciousness, a brief outline of the cognitive operations involved in the above planning stages follows. A brief definition of Generalized Empirical Method describes the focus of this process.

> Generalized empirical method operates on a combination of both the data of sense and the data of consciousness: it does not treat of objects without taking into account the corresponding operations of the subject; it does not treat of the subject's operations without taking into account the corresponding objects (Lonergan, 1985, p. 141).

What operations correspond with the planning stages in the above description of the process prior to the experiment? The pre-experimental tasks that need to be accomplished are initiated by questions towards planning and practicality. So, a series of questions such as, What kind of experiment will test the hypothesis? Who will write the research proposal? Who will draft the ethics consent form? Who will seek out human subjects? Who will reserve time on the MRI scanner? What type of recording devices is required? What information is to be communicated to the subjects prior to the conducting of the experiment? Who will draft a request for the medications required? One can easily notice that these are common sense activities orientated towards practicality and not scientific in a theoretical context. But these questions do reveal a reach for an ordered structure of achieving outcomes. The questions are eventually resolved through insights into the different talents or capabilities of the various colleagues involved in the planning stage. Once these insights are achieved the group makes decisions on who will do what. In this manner, an ordered structure, from the data of the hypothesis, to questions, to insights, to judgments about the correctness of their insights and to final decisions, occurs all through the planning stage. The cognitive structure orders the process heuristically towards achieving the desired outcome. I would note here that a parallel complexity occurs in venturing into work in any of the specialties, but brevity requires that I not repeat the details in the other specialties.

When these activities and tasks are accomplished, the next step is running the experiment, and even this process is not pure scientific knowing. It is in the order of doing in terms of implementing a process organized by all that has gone before, but it too requires planning. Ensuring all devices are ready and working properly, all the subjects are present, medications are prepared, initiating the proper efficient order and detailed operation of the experiment so that the data desired is objectively achieved. So, the operations involved are a checklist of questions, and in summary; Do we have everything here required to run this experiment? The question initiates an ordering of procedures of checking for last minute details. If all is ready, a judgment is made that all is ready, and a final decision is made to begin.

The overall outcome of all the work prior to running the experiment was to obtain data that will, hopefully, help in testing the original hypothesis. This entire process, then has been one focused on gathering data. It is a focus of attention towards that one outcome. Attention is an act of consciousness. As much as other operations such as questioning, understanding, judging and deciding occur throughout the process (Henman, 2013, p. 51. Lonergan, 1992, p. 343,361), attending to the outcome of gathering the data was the overall focus at that stage of the process. The final process in obtaining that data was to run the experiment.

The experiment was run as described above and in the article sited. It is in the order of doing where attention to detail of procedure is paramount in terms of each stage of the experiment. Timing of the various phases of the experiment is crucial. Earlier stages of planning were designed to ensure that such is the case. One can envisage possible test runs prior to the actual experiment to ensure that everyone involved understands his or her role or function and that all equipment is in proper working order. There are operations of cognition involved in attention to detail and one's understanding of not only his or her role, but also some overview of the entire experiment and how their function or role fits into the overall plan.

The experiment was conducted and the data had been recorded and reproduced.

The data consist of pictorials showing the results of the fMRI scans followed by two graphs all reproduced in the original article (Furey *et al*, 1997). Figure 1 of the fMRI scans reveals patterns of regional cerebral blood flow (rCBF) increasing during working memory (WM) tasks as compared with rest for the off and on drug conditions as well as the group comparison of activations. Figure 2 graphs the rCBF according to drug response in the right prefrontal cortex. Figure 3 consists of graphs showing the change in rCBF in relation to the changes in the reaction time of facial recognition. These three figures represent the data derived from the experiment.

Functional Interpretation

The second task is to understand just what this extended data means. This task is focused on interpretation or understanding. The division of tasks correlate to acts of cognition and not to different forms of data, which traditionally specified a science. First, the researchers attended to the many processes orientated to gathering data and now the second task is focused on understanding that data. The following is an excerpt from the article by the authors offering their interpretations of the data.

Given that acetylcholinesterase inhibitors prolong acetylcholine activity at the synapse, one might expect improvement in performance to be associated with increased rCBF in a task-specific brain region, yet we observed a reduction in right prefrontal rCBF. **One possible explanation is that reduced activation in right prefrontal cortex may reflect the shorter time required to perform the WM task,** although several papers, including a recent functional magnetic resonance imaging study, indicate that right prefrontal activity is associated more with maintenance of an active representation over the memory delay than with response selection.

The observed reduction of right prefrontal activity might have been a direct effect of physostigmine. Anatomically, the synaptic circuitry of cholinergic prefrontal fibers includes symmetric synapses on pyramidal cells; symmetric synaptic morphology is characteristic of inhibitory mechanisms. It is unclear, however, why a direct inhibition of right prefrontal activity would be correlated with improved WM performance. Moreover, direct inhibition demands neuronal activity and therefore may result in increased rather than decreased cerebral blood flow.

An alternative explanation for the physostigmine-induced reduction of the right prefrontal rCBF response is that right prefrontal activity is associated with the effort needed to perform the WM task,

and that this region is recruited as the effort required to perform the task increases. Prefrontal cortex may become more active with tasks that require a greater allocation of attentional resources. *A PET study of word list recall showed that increased memory load resulted in greater rCBF increases in frontal cortex as well as in other brain structures. Another PET study showed that increases in the difficulty of a face-matching task increased rCBF in the right midfrontal gyrus with a local maximum 6 mm from one of the anterior local maxima identified in this study. Similarly, studies of event-related electrophysiological potentials show that the response amplitude in the prefrontal cortex increases with task difficulty during WM.* **The anticholinesterase effects of physostigmine may enhance efficiency of WM processes, thus reducing the effort required to perform the task and the need to recruit prefrontal cortex.** The mechanism by which physostigmine reduces effortful processing during WM remains unclear. **Physostigmine may enhance efficiency by amplifying processing of information in the focus of attention[2] or by minimizing the effects of distracting stimuli** (Furey *et al*, 1997) (Bold and Italics formatting added)

The authors of this article have offered five possible explanations (Bold) for the results of the experiment and three alternative explanations (Italics) by other researchers conducting different experiments with a similar objective. The five explanations offered on Furey's experimental results were achieved by correlating the results of the three different recordings during the experiment. The researchers admit that much is still unclear about what exactly is enhancing the efficacy of working memory.

Functional History: Context of Interpretations

The fact that much is unclear and various possible interpretations have been offered brings us to a third task in the scientific process. This task focuses on the context of the experiment, the experimenter and the various

---

[2] The issue of attention or lack thereof may be the more adequate explanation of some forms of dementia and AD. Patterns of chemistry formed as correlates to experience are related to one's conscious attention to experience. Trauma, tragic loss, relational breakdown, and the inability to finalize situations on the level of decision can create a distraction so that one's ability to attend to experience is weakened resulting in insufficient patterns of chemistry correlating to the experience. As well as searching for pharmacological solutions to forms of dementia and AD, it would be helpful to work on both this solution as well as the therapeutic side and those innate systems of dysfunctionalism that exist in current social cultures. See Martin (2003) page 156–157 for a discussion of how repeating synapse form more permanent memories.

explanations with an effort to make a judgment on the context of interpretations. The specialist in this zone brings forth the various interpretations and judges which ones are viable for further study. In order to do this a review is carried out of the various stages leading up to the interpretations, as well as on the conducting of the experiment. Once such a reconstruction of the interpretative process is completed, the historian of contexts attempts to judge which interpretations should be brought forward for further study, further experimentation and reflection (Lonergan, 1972, p. 203). As much as the specialist, in this zone, is focused on judgment, all cognitive operations are active. Attention to the various stages of the process are required; understanding what the hypothesis is, what questions were asked, how the experiment was comprised, how it was run, how the data was collected and how it was interpreted. These stages of cognitional operations are all part of the process of arriving at a judgment on which interpretations are worthy of passing on to the next stage of the scientific process.

One needs a more than an adequate understanding of the science in question to function properly in this specialty. Lonergan makes the point in the following.

Clearly, therefore, the historian of any discipline has to have a thorough knowledge and understanding of the whole subject. And it is not enough that he understand it any way at all, but he must have a systematic understanding of it. For the precept, when applied to history, means that successive systems which have developed over a period of time have to be understood. The systematic understanding of a development ought to make use of an analogy with the development that takes place in the mind of the investigator who learns about the subject, and this interior development within the mind of the investigator ought to parallel the historical process by which the science itself developed (Lonergan, 1960, p. 130–132).

Functional Dialectics

The historical fact of having more than one possible interpretation of the data obtained from the experiment raises the issue of deciding which interpretation passed forward from functional history is the most viable or best explains the results of the experiment. The focus here at this stage is on the cognitive act of decision. How is such a decision to be accomplished? The differing interpretations offered reveal that much is still unknown about the workings and functioning of the brain in terms of working memory, the specific interaction of various chemicals involved, how brain chemistry affects attention: if so, how does it achieve this, and so much more? Obviously more experimentation will be required in the area of working memory. Are there particular points made in the interpretations that can be brought forward that would appear to have some basis for further development in understanding the functioning of working memory? This is addressed in the discussion section of the essay. This process involves comparison, but not in some common sense descriptive mode. Such comparison requires that the dialectician operate out of a foundational context regarding an understanding of his or her own mental operations as well as familiarity with the rules of interpretation[3] that contribute to an eventual genetic sequence of development (Lonergan, 1992, pp. 600–603). The task of dialectic is to establish "a final objectification of horizon when the results of the foregoing process are themselves regarded as materials, when they are assembled, completed, compared, reduced, classified, and selected, when positions and counter-positions are distinguished, when positions are developed and counter-positions are reversed (Lonergan, 1972, p. 250)."

Interlude

Before moving forward to further stages, it will be helpful to outline briefly, what has been going forward up to this point in the process. Four tasks have been described 1) research with a focus on attention to gathering data, 2) interpretation with a focus on understanding the data, 3) history with a focus on judging the contexts of the various interpretations offered and 4) dialectic, deciding which interpretation(s) is worthy of further study. Each stage reflects on the results of the former stage. The results of each stage are the "datum" for the next stage, but the focus of each stage is on a different mental operation. So, researchers focus on attention, interpreters on understanding, historians on judging the story and dialecticians deciding about the progress of the story. The order of the focus parallels the unfolding structure of human knowing (Henman, 2013, p. 51, Lonergan, 1992, p. 299). Specialization is grounded in the cognitive operations. Because it follows the cognitive order, from experience (attending to data), understanding, judging and deciding, it provides a greater possibility of more cumulative and progressive results. Such interdependence of the emerging specializations brings to the process a form of collaboration that is functional in as much as each specialist is intentionally focusing on providing data for the next task.

---

[3] The rules of interpretation would require a further lengthy article. See Lonergan, 1992, pp. 600–603.

The results of the first phase provide data for the second phase. A further feature of these four specialties is that they bring forward work from the past into the present with an orientation to the future. The second phase begins with the specialty foundations but springs from the dialectician's efforts to point towards progress.

Functional Foundations

This second phase of functional collaboration, as an orientation to the future, pushes forward the decisions of the previous specialty on what specifically is a position from phase one. Foundations, the fifth specialty, is focused on decision as is dialectic. Dialectic determined which explanation was the best available from those offered by the historian. Foundations focus on whether or not the interpretation offered is an adequate explanatory account of the data. Are there results from the previous work that will lead to further developments in the field of neuroscience research relating to medications to improve working memory and treat Alzheimer's disease? This stage of deliberation is one of appropriation of the researcher's own procedural cognitional operations that underlie the position. The following quotation from the discussion section of their article states what the researchers considered as the main result of their experiment.

> The results of this study show that enhancement of cholinergic neurotransmission results in an improvement in WM efficiency that is correlated with an alteration of brain activity in a cortical region known to play a central role in WM (Furey *et al*, 1997).

The specialty foundations, takes a position on the decision offered in dialectics, but what is a position? Lonergan offers the following as the cognitional and the epistemological foundations of a position. The decision offered in dialectic will be a position if:

(1) the knowledge reached is the result of the process of intelligent and rational operations and not some subdivision of the 'already out there now';

(2) the researcher becomes known when he or she affirms him/herself intelligently and reasonably and is not known yet in any prior 'existential' state; and

(3) objectivity is conceived as a consequence of intelligent inquiry and critical reflection, and not as a property of vital anticipation, extroversion, and satisfaction (Lonergan, 1992, p. 413).[4]

Foundations, then, decides if the account offered is an explanatory account of the data correlates with the three characteristics of a metascientific position. The above quotation from Furey is a descriptive account. A partial explanation for this result is that physostigmine which is an acetylcholinesterase inhibitor increases the duration of action of acetylcholine at the synapse. Regional blood flows in the prefrontal cortex reacted differently than expected, and the authors were unable to explain this and offered various possible explanations. No definitive explanation was offered in the article. Underlying this failure is methodological unclarity. Foundations, then focuses on what methodological improvement in a given research experiment is a position that is worth passing on. Resolving the lack of a methodological clear explanation serves as a position to be passed on to our next specialty, policies.

Functional Policy

The task of metapolicy is to work out what plans can be constructed or developed grounded in the position taken in foundations. The focus is on judgment, to develop policies and to judge which policies are viable. The function of developing policies is to establish a sequence of plans of operation that are orientated towards further development within the science. What improved plans of operation can be developed based on the results of Furey's experiment? The results of Furey's experiment reveal that further experimentation is required especially as some results were unexpected regarding the reduction of rCBF in the right prefrontal region. What is the methodological explanation for this and how is it relevant to working memory? A policy to be established is that further precise experimentation, regarding the correlation between changes in brain activity and the enhancement of cholinergic neurotransmission be carried out in an effort to understand the unexpected results. We now move to our next specialty, systematics, which focuses on understanding what this policy means in the full context of a genetic systematics of the science.

---

[4] I have edited the quotation of Lonergan to fit the particular context of neuroscience research.

Functional Systematics

Systematics focuses on the cognitive act of understanding. The specialist is attempting to understand contextually the results of the prior specialties that have culminated into policies. The overall outcome is to provide a genetic systematic understanding of the policy that can be developed into an adequate form of communication for both all other seven specialists and for the various possible audiences. What does the policy of further experimentation mean? On what forms of experimentation should researchers focus? What methodological steps, if any, have been missed or operationally misplaced? These questions are a small sampling of the types of questions that systematics attempts to answer in relationship to the particular experiment discussed in this article. But to conceive of it in its intrinsic structure is a huge task of the future. There is at present little reach towards even an imagining of the layered structuring of its developmental neurodynamics. Think of the problem of accounting for the evolving patterns of the prenatal brain; think of the growth of a sunflower. Then think of an inquiry into such different thinkings that would become a genetic sequencing of the open advancing of theories of all such things.

Functional Communications

The understanding that has come forward from systematics must now be expressed selectively in various ways that are appropriate to the various possible audiences. Those audiences certainly include researchers, clinicians and non-technical. Furey's article was written for other researchers. The language is technical, and one requires a background in neuroscience in order to understand the meaning correctly. Clinicians would have some grasp of the article's meaning but an adequate form of communication for the clinician would focus more on possible treatment regimens. The patient or the non-technically trained person would require another form of expression that would provide that group with some common sense description of what is going forward. At this stage, Furey's work would not warrant any treatment regimens.

For our purposes the first form of communication is relevant, that of the technical form of expression to other experimental neuroscientists. This form of communication is now data to be recycled through the eight functional specialties. Note that the communication also is to reach the various specialists. In other words, the process of functional specialization begins again, but with a fresh lift of method and context.

The results of Furey's work became data for further experimentation. In fact, many experiments were carried out in the years following Furey's work. F. Coelho and J. Birks carried out a study of experiments and reviews by searching the Cochrane Controlled Trials Register. One could detect elements of our precise collaborative process in this work, but it would be a separate article. The overall conclusion was that the evidence of effectiveness of physostigmine for the symptomatic treatment of Alzheimer's disease is limited (Coelho, Birks, 2001). This result differs from the conclusion of Furey's 1997 article.

> The results of this study show that enhancement of cholinergic neurotransmission results in an improvement in WM efficiency that is correlated with an alteration of brain activity in a cortical region known to play a central role in WM (Furey *et al*, 1997*)*.

It is to be noted that Furey's experiment focused on WM in healthy subjects whereas Coehlo and Birk's search through experiments and reviews focused on experiments with AD subjects. The interpretation of these results led to other experimentation with the cholinergic drugs donepezil, rivastigmine and galantamine in order to assess their possible aid in the treatment of AD. These reviews can be viewed at the Pub Med, US National Library of Medicine at: http://community.cochrane.org/sites/default/files/uploads/EPPR/CDSR_Impact-factor_and-usage-report_2013.pdf or the Cochrane Library Database Systems at http://www.ncbi.nlm.nih.gov/pubmed/16437532.

The above search by Coelho and Birk is classified as research, the first specialty, a gathering of statistical data to be cycled through the eight functional specialties. In that manner of organizing tasks, a higher control of understanding is initiated.

Dr. Furey has continued with the National Institute of Mental Health (NIMH) to experiment with working memory and various studies have been published of her results. See https://intramural.nimh.nih.gov/research/clinicians/sc_furey_m.html for some of her publications and more recently publications with Science Direct (Furey, 2009, pp. 322–332). Furey's work influenced the focus and expansion of her later research as well as others. In that manner, there was ongoing collaboration and extension of the original work into other possible zones. What that process lacked was a division of labour structured by the order of the operations of human cognition, as outlined in this article above, but the components of that possibility were there.

Functional Specialization (Lonergan, 1972, ch.5, Shoppa, Zanardi, 2014, p. 28, McShane, 2013)

A brief outline of the different specialties and their functions has been provided above. The following provides a brief overview of the structure and meaning of this outline. The full work of research from data to results has been divided up into eight tasks grounded in the cognitive operations of the human intellect. Why are there eight divisions? There are four distinct operations of the intellectual process: attention, understanding, judgment and decision. These four operations give a heuristic pattern to knowing. Each operation presupposes the former operation as well as reaching for the succeeding operation. Data evoke curiosity expressed in the form of a question. A question seeks insight or understanding, understanding requires formulation into a judgment and judgment occurs as an answer to an IS question; Is it so? A decision is in the form of a to-do Yes or No regarding the method of verification. This heuristic is developmental as former verified insights into data form the ground for further questions and insights to be added to the composite of knowledge in a particular zone.

Each functional specialty is grounded in and focused on one of the cognitive operations so that development is made possible in an organized manner. There is a methodological ordering of content of the science in question (McShane, 1970, p.259). The first four specialties, the first phase, bring forward work from the past, and the second phase is orientated towards the future as to what this former work can contribute to the development of a particular zone or science. The diagram on the following page provides an image of the flow of functional specialization.

Phase One–Past   **Phase Two–Future**

Dialectics	*Decision*	Foundations
History	*Judgment*	Policy
Interpretation	*Understanding*	Systematics
Research	*Attention*	Communications

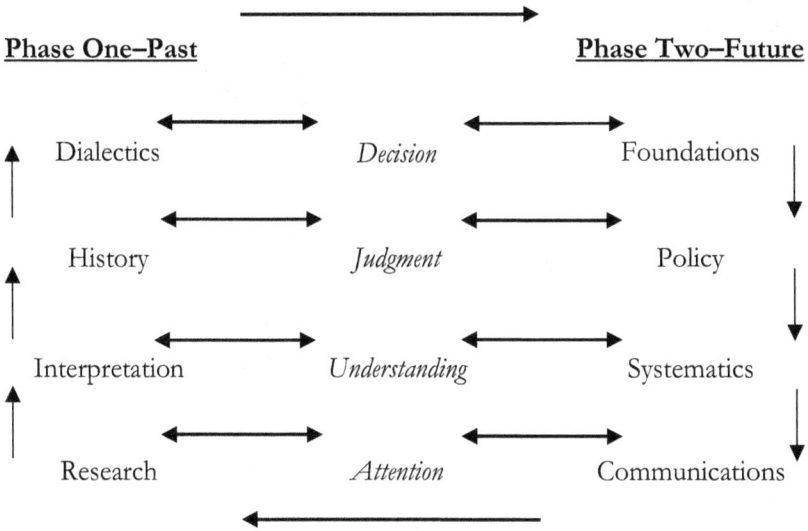

A brief summary of the eight specialties describes the specific focus of each specialty (Anderson, 1996, p. 167).

Research: Collecting and selecting the relevant data

Interpretation: Establishing the meaning of the data.

History: Working out what is going forward.

Dialectics: Sorting through the various interpretations and histories with the aim of deciding which one is the best explanation.

Foundations: Expressing the best directions forward in a way that is not tied to particular places, ages and times.

Policy: Reaching for relevant pragmatic truths within a foundational context.

Systematics: Drawing on past strategies and discoveries while envisaging future concrete possibilities and their probabilities.

Communications: Collaborative reflection on the local level that selects creatively from the range of possibilities developed in the prior seven specialties with a view to developing forms of communication to the various possible audiences.

Discussion

In the Introduction, I quoted two definitions by Bernard Lonergan and raised two questions, which I repeat here and attempt to answer. Method is a framework for collaborative creativity (Lonergan, 1973, p. xi) and method is a normative pattern of recurrent and related operations yielding cumulative and progressive results (Lonergan, 1973, p 4).

In as much as each of the specialties is grounded in the heuristic structure of cognition there is also a functional capacity of performance imbedded in the structure. Each specialty has a function to be passed on to the next specialty and those functions form an interlocking of interdependence that works toward a completed whole. The division of labour that functional specialization orchestrates also provides a form of collaboration that increases the probabilities of cumulative and progressive results in a science. No scientist today can be a specialist in all zones of a science (Zanardi, 2011, p. 19–54).[5] Neuroscience draws on physics for its understanding of electrical voltage regarding synaptic events, chemistry and biochemistry, cellular biology, anatomy, physiology, and psychology in its attempts to understand the human brain and its relationship to the body, to development and to consciousness.

For good or ill a division of labour has to be accepted, and this is brought about by dividing and then subdividing the field of relevant data. ...to make the specialist one who knows more and more about less and less (Lonergan, 1973, p. 125, Agoritsas, 2013, p. 448).[6]

A division of labour focuses not just on the data of the content results of each specialty but also on the data of the cognitive operations making possible a methodical ordering of content which makes possible a higher probability of cumulative and progressive results in an age of ever increasing data. Such a division distributes the various tasks in an orderly and manageable manner. The focus on both the data passed on from one

---

[5] Professor Zanardi explores methodologically the binding problem and the difficulty of working out the various correlations and processes involved in the formation of sensation. The case for functional specialization is supported by his analysis of what is increasingly found to be a series of very complex processes.
[6] Agoritsas and Guyatt report that Medline publishes more than 2000 articles a day. A portion of those articles would be in the area of neuroscience and others related to the ongoing work of neuroscience. This does not include textbooks. This immensity of output only serves to further support the need for a method of organizing the content of all the sciences if development and progress are to be achieved and the probabilities of such are to be raised.

specialty to another and on a particular cognitional operation challenges the scientist not only to be a specialist in one zone of a science, such as the influence of drugs on working memory, but also as a specialist in one of the functional specialties. So, the specialist focused on history need not only an adequate background on working memory as it pertains to neuroscience but a good knowledge of the history of work accomplished on working memory and an understanding of the role of judgment in scientific knowing.

Would a normative pattern of related operations be a framework for collaborative creativity and furthermore, would such activity provide cumulative and progressive results? Creativity is the exercise of the human capacity to raise the questions that have not been asked and then to follow through to achieve the insights that call for verification. There is the challenge of openness in the scientist's psychological and intellectual poise towards their work (McShane, 2013, chs. 3 & 4). Openness is expressed in the form of one's curiosity, one's ability to ask questions, uninhibited by ulterior motives. "Arts and our aesthetic experience in the form of interpretation are thus connected to how we learn (Gillis, 2015, p. 138)."[7] Curiosity is the genesis of one's questions leading to understanding, the normal and natural resolution to a question (Henman, 1984, ch. 1 & 2. Lonergan, 1992, pp.34–35). Any other motive for asking a question becomes an inhibition to scientific achievement (Lonergan, 1992, Ch. 6, sec. 2.7 & ch. 7 sec 6, 7, 8).

There are three distinct forms of questions that are exercised during the functioning of the eight functional specialties. There is the "what" question in functional research: what data are relevant to this study? There is the What question for interpreters: what is the meaning of this datum? The historian: what is the context of this interpretation? The dialectician: what interpretation is the best? The Foundational person shifts to a new form of question: an Is it so? type question or Is this interpretation a position? Policy asks the question: What is to be done with this interpretation? Systematics asks what this policy means. Communications asks what form of communication is to be developed and for whom. Therefore, we have What questions, Is questions and What to do questions.

Questions can have many motivations or distractions. A question normatively seeks understanding. Functional specialization will begin to

---

[7] Gillis explores the aesthetic experience in terms of its ability to develop the openness of one's wonder as the ground of human aspiration and well-being in a world in which artistic training and scientific training can and often does stifle one's curiosity. Gillis was kind enough to allow me to read her completed thesis and she does have plans for eventual publishing.

offset ulterior motivations by its focus on cognitive operations. Such extended focus over time would achieve this outcome. That achievement would increase the creative aspect of curiosity with a more focused attention on the desire to understand as opposed to other motivations (Henman, 2010, 2012). Because the pattern of functional specialization divides the question types and cognitive operations among specialists, it encourages and shares the creative capacity among the specialists increasing the probability of development. This sharing of intellectual capacities further reveals the collaborative character of functional specialization.

Cumulative and progressive results occur when labour is divided along the lines of the cognitive operations because the heuristic structure of attention, understanding, judgment and decision are the ground of human progress (Lonergan, 1992, p. 8). Insights correct, revise or build on former insights. Insight is an integral operation. The centrality of insight in cognitive activity provides the possibility for development in a science or zone of enquiry. The act of insight integrates former insights into a higher viewpoint modestly exemplified in the transition from understanding addition to understanding subtraction or the transition from Newtonian mechanics to Relativity theory. The emergence of new questions and new insights is not an automatic process, but the order is intelligible. They are the result of a dedication to understand, or the unrestricted desire to understand (Lonergan, 1992, p. 667–69), a curiosity that in principle, knows no bounds.

Conclusion

The operation of functional specialization in the sciences would eventually implement Generalized Empirical Method (GEM) constituting the New Science.[8] Its implementation would ensure that scientists take into account

---

[8] Bernard Lonergan outlined functional specialization as a method of establishing theology as a science. It has been pointed out by Professor Karl Rahner that the method would also apply to any of the sciences. Rahner, K. (1970) Kritische bemerkungen zu B.J.F.Lonergan's Aufsatz: Functional specialties in theology, Gregorianum, vol. 51, page 537. "Lonergan's theological methodology seems to so general that it applies equally to all sciences, and so is not a method of theology as such but a general method of science illustrated by examples from theology." Quinn (2015) explores how generalized empirical method within the context of functional collaboration will respond to the problems of imagined molecules in systems theory in biology. Pages 18–19. Shute (2015) listed the following applications of functional specialization in footnote 8, page 69.

not only the data of sense but also the data of consciousness when performing scientific work thereby increasing the odds of a systematic understanding of data generating a genetic sequencing of instances of development. As the eight different specialists become more at home with GEM (McShane, 1976) any tendencies towards reductionism will be gradually countered (Henman, 2015). Only through the ordered functioning of the specialties, will a more adequate understanding of their function as the new science come to be appreciated in terms of the functional ordering of results that occur. Without knowledge of the order of the operations expressed in functional specialization, a degree of haphazardness is present in the work of a science. [9] Functional specialization can reduce such randomness due to the very structure of the human intellectual operations increasing the probabilities of cumulative and progressive results. The

---

"Philip McShane has explored the implications of functional specialization in physics in "Elevating *Insight*: Spacetime as Paradigm Problem," Method: Journal of Lonergan Studies 19/2 (2001); in economics in, *Economics for Everyone* (Halifax: Axial Press, 2002), chapter 3; in linguistics in *A Brief History of Time* (Halifax: Axial Press, 1998); in musicology in *The Shaping of Foundations* Washington, DC: University Press of America, 1976), chapter 2, available online at www.philipmcshane.org; and in literature in *Lonergan's Challenge to the University and the Economy* (Washington, DC: University Press of America, 1980), chapter 5 available online at www.philipmcshane.org. Bruce Anderson in *Discovery in Legal Decision-Making* (London: Kluwer Academic Publishers, 1996), discusses functional specialization as applied in philosophy of law. Terrance Quinn discusses functional specialization in mathematics in "Reflection on Progress in Mathematics," *Journal of Macrodynamic Analysis* 3 (2003) 97–116. Ian Brodie has applied functional specialization to religious studies in "Bernard Lonergan's Method and Religious Studies: Functional Specialties and the Academic Study of Religion," M.A. thesis, Memorial University, 2001. More recently there is Scott Halse, *Functional Specialization and Religious Diversity: Bernard Lonergan's Methodology and the Philosophy of Religion*, PhD. Thesis, McGill University, 2008. Most recently, a functional specialist approach has been applied to the practical issue of housing. See Sean McNelis, *Making Progress in Housing: A Framework for Collaborative Research* (Oxon: Routledge, 2014)."
[9] Theil, Stefan, (2015) Scientific American, Volume 313, Issue 4. Theil highlights the difficulty of collaboration between the European Human Brain Project (HBP) and the American Brain Initiative. Politics and funding are part of the challenge in this particular case but a more overarching and profound challenge is the establishment of a method that correlates the two projects and their desired outcomes intelligently. For the entire article see:
http://www.scientificamerican.com/article/why-the-human-brain-project-went-wrong-and-how-to-fix-it/?WT.mc_id=SA_MB_20150923

transition from present operations to GEM, taking into account both the data of sense and the data of consciousness, will be one of historical relevance and far-reaching in terms of time. It will be one of centuries before we are at home in such a process.

References

Agoritsas T., Guyatt G., (2013) "Evidence-based Medicine 20 Years on: A View from the Inside", Can J Neurol Sc. 40: 448–449.

Anderson, B. (1996) *Discovery in Legal Decision-making*, Dordrecht: Kluwer Academic Publishers.

Douglas, H. (2013) "Pure Science and the Problem of Progress", *Studies in History and Philosophy of Science* (46) pp. 55–63.

Furey, M. (2009) "Cholinergic Modulation of Visual Working Memory during Aging: A Parametric PET Study", *Brain Research Bulletin*, Vol. 79, Issue 5.

Furey, M., Pietrini, P., Haxby, J. (2000) "Cholinergic Enhancement and Increased Selectivity of Perceptual Processing during Working Memory". *Science* 290.

Furey, M., Pietrini, P., Haxby, J., Alexander, G., Lee, H., VanMeter, J., Grady, C., Shetty, U., Rapoport, S., Schapiro, M., and Freo, U. (1997) "Cholinergic Stimulation alters Performance and Task-specific Regional Cerebral Blood Flow during Working Memory". http://www.pnas.org/content/94/12/6512.full

Gillis, A. (2015) *Aesthetic Dimensions of Education: Exploring a Philosophical Pedagogy using Dialogue with Arts Learners and Educators*, Unpublished PhD thesis at Simon Fraser University, Vancouver, BC.

Henman, R. (2015) "Generalized Empirical Method: A Context for a Discussion of Language Usage in Neuroscience", http://www.crossingdialogues.com/current_issue.htm Vol. 8, Issue 1. Dialogues in Philosophy: Mental and Neuro Sciences, Rome, Italy.

Henman, R. (2013) "Can Brain Scanning and Imaging Techniques Contribute to a Theory of Thinking?" http://www.crossingdialogues.com/issue22013.htm Vol. 6, Issue 2. Dialogues in Philosophy: Mental and Neuro Sciences, Rome, Italy.

Henman, R. (2012) "An Ethics of Philosophic Work", Journal of Macrodynamic Analysis, Vol. 7, pp. 44–53, http://journals.library.mun.ca/ojs/index.php/jmda/article/view/359

Henman, R. (2010) "Teaching Foundations in Peace Studies", Journal of Macrodynamic Analysis, Vol. 5, pp. 20–29, http://journals.library.mun.ca/ojs/index.php/jmda/article/view/179/124

Henman, R. (1984) *The Child as Quest*, University Press of America, Maryland, USA.

Lonergan, B. (1992) *Insight: A Study of Human Understanding*, (CWL 3) University of Toronto Press, Toronto, Canada.

Lonergan, B. (1972) *Method in Theology*, Darton, Longman & Todd, UK.

Lonergan, B. (1960) *De Intellectu et Methodo*, Lectures, Gregorian, 1959, Translated by Michael Shields as *Understanding and Method*.

Martin, G. (2003) *Essential Biological Psychology*, Oxford University Press, NY, USA

McShane, P. (2013) *Futurology Express*, Axial Publishing, Vancouver, BC.

McShane, P. (1976) *The Shaping of Foundations: Being at Home in the Transcendental Method*, University Press of America, Wash., D.C..

McShane, P. (1970) *Randomness, Statistics and Emergence*, Notre Dame Press, Ind. USA.

Quinn, T. (2015) "Reaching for Collaboration in *Insight* (and Beyond)", Journal of Macrodynamic Analysis, Vol. 8. http://journals.library.mun.ca/ojs/index.php/jmda/article/view/1631/1224

Rahner, K. (1970) "Kritische Bemerkungen zu B.J.F.Lonergan's Aufsatz: Functional Specialties in Theology", Gregorianum, vol. 51.

Robbins, T., Mehta, M., and Sahakian, B. (2000) "Boosting Working Memory", *Science* 290.

Ross, W.D. (1949) *Aristotle's Prior and Posterior Analytics*, Oxford, UK.

Shoppa, C., Zanardi, W. (2014) *Cracking the Case: Exercises in the New Comparative Interpretation*, Forty Acres Press, Austin, Texas.

Shute, M. (2015) "Functional Collaboration as the Implementation of Lonergan's Method, Part 1: For What Problem is Functional Collaboration the Solution?" Journal of Macrodynamic Analysis, Vol. 8. http://journals.library.mun.ca/ojs/index.php/jmda/article/view/1625

Zanardi, W. (2011) "Preconceptual Apprehension and Evaluation of Objects", Unpublished, http://www.philipmcshane.org/wp-content/themes/philip/online_publications/series/fuse/fuse-12.pdf

At the conclusion of my Introduction I mentioned that the focus of this Epilogue was to be on the first word of the third chapter: *Implementing*. An earlier ambition would have had me enlarge on the contexts of relevance of the message of the three chapters, extending McShane's pointing of the Foreword. But it seems most appropriate now—a strategy of effective implementation—to home in on the problem of realizing, in seminal fashion, the components of implementation. That there are separate and distinct components of the intended collaboration are, no doubt, fairly obvious. One can think of these in terms of the exchanges—the image of baton-exchange in relay racing, which Phil McShane often uses, comes to mind—at the fronts of each specialization of collaboration.[1] Researchers pass the baton to interpreters, who pass the more refined baton to historians, etc. etc. The refinements pertain to possible improvements in the pragmatic viewpoint of global well-being, whatever the particular discipline or zone of inquiry that is involved.[2] In the zone of my attention in this book there are to be concrete improvements, obviously, in the character and dynamic of work in what may be called, but it is dated, indeed, obsolete, in the present context, pure neuroscience. But the outreach goes to every facet of applied neuroscience: the neurodynamics of persuasion, the psychology of monetary exchanges in shops and stocks, the psychodynamics of the structuring of books and buildings relating to education, from kindergarten to graduate seminars.

The incipient listing of the etc. etc. in the previous paragraph might have led you to a count, indeed to a count of 8 generic types of exchange, reaching the eight in the final output of the field of communications, the eighth specialty to what I might call the shaping of daily practicalities. Is there another zone of baton-exchange involved? Here it is important to take note of the peculiarity of the **idea** of functional collaboration. It is an idea that is, potentially, a full heuristic, indeed an idea that is to blossom into a full

---

[1] See Philip McShane, "The Importance of Rescuing Insight," *The Importance of Insight. Essays in Honour of Michael Vertin*, edited by J.J. Liptay and D.S. Liptay, University of Toronto Press, 2007, 202ff.

[2] This involves a complex heuristic of the concrete interweaving of disciplines. So, one finds that disciplines converge in interdependence as they move from research to dialectic and then diverge, but freshly integrated, in their move to concrete application. A key point to note is how they increasingly come to share common foundations.

pragmatic metaphysic.[3] A full pragmatic effective metaphysics would entail, weave into thoroughly, a feedback dynamic of self-correction and self-improvement.

The feedback needs to be conceived more and more adequately as the ethos, the collaborative methodology, takes hold. There is, of course, the internal set of feedback structures, which can be symbolized by the 8-by-8 matrix C_{ij} that refers to exchanges between specialties, but here I am interested in this ninth zone, which ranges from outputs of the last specialty to inputs to the first specialty.[4]

I would note that this is a zone that has been given little attention so far by those advocating the global collaboration that is my topic. Bernard Lonergan can be seen to explicitly discuss the matter of implementation only twice: a first effort in his work *Insight*, where he writes in terms of a sort of logic of achievement.[5] His later and final effort, too, is unsatisfactory: a final chapter of his book *Method in Theology* where he writes of his concern about Communications.

> It is a major concern, for it is in this final stage that theological reflection bears fruit. Without the first seven stages, of course, there is no fruit to be borne. But without the last the first seven are in vain, for they fail to mature.[6]

McShane pushes the topics of Implementation and Communications towards a much fuller conception of their heuristics, but he does not tackle the extra-specialist zone to which I am drawing attention here. That zone's study might be considered as the psychosociology of everyday persuasion and innovation. But note that the study required, paradoxically, is the study that is the eightfold collaboration that is the topic of chapter three, and the core of both its successful inner (think of the work within the specialties) and out-

---

[3] The process touched on in the previous note is to slowly lift Lonergan's original definition of metaphysics, especially in regard to implementation, into a quite novel pragmatic effectiveness. It is worthwhile to repeat that definition here: "Explicit metaphysics is the conception, affirmation, and implementation of the integral heuristic structure of proportionate being." *Insight*, *CWL* 3, 416.

[4] A context for this searching and diagramming is "Communications and Metaphysics as Science," chapter 16 of Philip McShane, *The Allure of the Compelling Genius of History*, Axial Publishing, 2015.

[5] See *Insight*, 424-25.

[6] *Method in Theology*, 355.

reaching (think of the zone of our present interest) dynamic is the topic of the previous two chapters.

What is the character of the paradox involved here? One finds it by investigating, in prescientific fashion, the character of everyday persuasion and innovation. And in that investigation there is the solitude of such prescientific investigation. Characters of the study are rare, and seeds of the characteristics of desire are scarce.

Here you may notice a twist in my reflections that brings us back to the distinctiveness of this book. Is it prescientific? Certainly it is, as a communication. It is in the zone that I can name here as "a ninth zone."[7] It addresses people who, despite qualifications in empirical psychology and its applications, are not beyond the world of present commonsense views of persuasion and innovation. Indeed, they are so entrapped in that common sense that the message of all three chapters is learnedly, if casually, shrunken to an ideological message, a utopian proposal, especially since the shrinking cuts off the discomfort and dread that goes with a paradigm shift of complex global proportions.[8] There is a massive and solidly established existential gap that holds the neurodynamics of those studying neurodynamics in firm patterns of conventional self-neglect.

How does one break forward from such monstrous conventions, conventions that ground the behavior patterns of greedy or gracious poise that make our lives unlivable, our sciences inadequate, our educational structures brutalizing? The problem brings to mind Lonergan's citing of William James description of three stages in the career of a theory. First, "it is attacked as absurd; then it is admitted to be true, but obvious and insignificant; finally it is seen to be so important that is adversaries claim that they themselves discovered it."[9] In that same essay he begins his concluding paragraph with the question, "Is my proposal utopian?", and goes on to repeat James' view. But what might I say of that view here in relation to that theory proposed in this book?

---

[7] The ninth zone is the area of common sense, influenced by functional collaboration's output. And, providing input to the ongoing collaboration. See Philip McShane, *The Allure of the Compelling Genius of History*, Axial Publishing, 2015, 188-190.

[8] On the topic of dread and the cultural and existential gaps that haunt progress, see Lonergan, *Phenomenology and Logic*, CWL 18, 280–97.

[9] William James, *Pragmatism*, London, Longmans, 1912, 198. Quoted in Lonergan's essay, "Healing and Creating in History", pp. 100–109 of *A Third Collection: Papers by Bernard Lonergan*, Edited by F.E. Crowe, Paulist Press, 1985, p. 102.

Could it be that the adversaries that come to claim my view for themselves are to be the neuroscientists themselves? Well, maybe not, since my hope is that my introduction to the New Science will seed a little doubt in some few minds working in the area, so that the claim of inventing would be shiningly suspect. Still, they could legitimately claim that they had uncovered it themselves. Indeed, that uncovering is not unlikely, for their science is nudging them at present to uncover themselves, a discovery that will nudge them to review with horror a history of missing the point of the selves to be discovered, the selves that makes human neurodynamics so radically different from that of the apes. The genitive there is both objective and subjective, and so we can end with a liberating laugh. Apes do not have neurodynamic theories.